I0094937

Embodied Memories, Embedded Healing

Environment and Society

Series Editor: Douglas Vakoch

As scholars examine the environmental challenges facing humanity, they increasingly recognize that solutions require a focus on the human causes and consequences of these threats, and not merely a focus on the scientific and technical issues. To meet this need, the Environment and Society series explores a broad range of topics in environmental studies from the perspectives of the social sciences and humanities. Books in this series help the reader understand contemporary environmental concerns, while offering concrete steps to address these problems.

Books in this series include both monographs and edited volumes that are grounded in the realities of ecological issues identified by the natural sciences. Our authors and contributors come from disciplines including but not limited to anthropology, architecture, area studies, communication studies, economics, ethics, gender studies, geography, history, law, pedagogy, philosophy, political science, psychology, religious studies, sociology, and theology. To foster a constructive dialogue between these researchers and environmental scientists, the Environment and Society series publishes work that is relevant to those engaged in environmental studies, while also being of interest to scholars from the author's primary discipline.

Recent Titles in the series

Embodied Memories, Embedded Healing: New Ecological Perspectives from East Asia, edited by Xinmin Liu and Peter I-min Huang

Wetlands and Western Cultures: Denigration to Conservation, by Rod Giblett

Sustainable Engineering for Life Tomorrow, edited by Jacqueline A. Stagner and David S. K. Ting

Nuclear Weapons and the Environment: An Ecological Case for Nonproliferation, by John Perry

Portland's Good Life: Sustainability and Hope in an American City, by R. Bruce Stephenson

Conservation, Sustainability, and Environmental Justice in India, edited by Alok Gupta

Environmental and Animal Abuse Denial: Averting Our Gaze, edited by Tomaž Grušovnik, Reingard Spannring, and Karen Lykke Syse

Living Deep Ecology: A Bioregional Journey by Bill Devall, edited with an introduction by Sing C. Chew

Environmental and Animal Abuse Denial: Averting Our Gaze edited by Tomaž Grušovnik, Reingard Spannring, and Karen Lykke Syse

Secular Discourse on Sin in the Anthropocene: What's Wrong with the World? by Ernst M. Conradie

Embodied Memories, Embedded Healing

New Ecological Perspectives from East Asia

Edited by
Xinmin Liu and Peter I-min Huang

LEXINGTON BOOKS
Lanham • Boulder • New York • London

Published by Lexington Books
An imprint of The Rowman & Littlefield Publishing Group, Inc.
4501 Forbes Boulevard, Suite 200, Lanham, Maryland 20706
www.rowman.com

6 Tinworth Street, London SE11 5AL, United Kingdom

Copyright © 2021 The Rowman & Littlefield Publishing Group, Inc.

All rights reserved. No part of this book may be reproduced in any form or by any electronic or mechanical means, including information storage and retrieval systems, without written permission from the publisher, except by a reviewer who may quote passages in a review.

British Library Cataloguing in Publication Information Available

Library of Congress Cataloging-in-Publication Data

Names: Liu, Xinmin, 1952- editor. | I-min Huang, Peter, 1953- editor.
Title: Embodied memories, embedded healing : new ecological perspectives from East Asia / edited by Xinmin Liu and Peter I-min Huang.
Description: Lanham : Lexington Books, [2021] | Series: Environment and Society | Includes bibliographical references and index.
Identifiers: LCCN 2021036060 (print) | LCCN 2021036061 (ebook) | ISBN 9781793647597 (Cloth) | ISBN 9781793647603 (eBook)
Subjects: LCSH: Ecocriticism—East Asian. | Ecology—Philosophy. | Philosophy, East Asian.
Classification: LCC PN98.E36 E543 2021 (print) | LCC PN98.E36 (ebook) | DDC 809/.9336—dc23
LC record available at https://lccn.loc.gov/2021036060
LC ebook record available at https://lccn.loc.gov/2021036061

Contents

Foreword

Scott Slovic

In the final pages of his classic work of nature writing, *Arctic Dreams*, American author Barry Lopez recounts a visit to the remote tip of St. Lawrence Island, on the edge of the Bering Sea. Gazing out at the vast expanse of ice and water, the author writes at length about his bow of respect to the place itself. "I held that bow," he explains, "until my back ached, and my mind was emptied of its categories and designs, its plans and speculations. I bowed before the simple evidence of the moment in my life in a tangible place on earth that was beautiful" (414). This bow of heartfelt respect—something fellow nature writer Richard K. Nelson refers to as a "gesture of reverence" (qtd in Sherman Paul, "Making the Turn" 377)—is an adaptation of a common East Asian posture, typically used in greeting or bidding farewell to another human being. For Lopez, the bow is used both symbolically and as an embodied manifestation, on the penultimate page of his masterpiece, to indicate utmost "appreciation" (415) for the more-than-human world.

When he received the National Book Award for *Arctic Dreams* in November 1986, Lopez invoked another idea from East Asia to describe his central ethos as a writer, the Japanese principle of kotodama. This term, he explained in his acceptance speech, "refers to the spiritual nature, the soul, if you will, of a word. What it means for writers is you have a responsibility that goes beyond yourself to care for the spiritual quality, the holy quality, the serious quality of language" (Keefer, "Lopez's Book on Arctic" 1B). Similarly, Lopez used the concept of the "bow" in referring to the writer's subject matter, not only to language itself. As he told interviewer Nicholas O'Connell, "The bow of respect toward the material means try to understand what's coming from it, not what you are trying to impose on it. Listen. Pay attention. Do your research. Try to learn. Don't presume" (16).

Two days following the death of Barry Lopez on Christmas Day 2020, as I write this Foreword, it seems fitting, even necessary to mention how this icon of American environmental literature relied deeply on cultural adaptations from East Asia in developing his personal beliefs about writing and about the psychological attachment between people and the physical world. This fascination with various ecological concepts from East Asia is the central focus of *Embodied Memories, Embedded Healing: New Ecological Perspectives from East Asia.*

The connection between American environmental literature and East Asia is pervasive and has strong roots. Many scholars have noted Henry David Thoreau's indebtedness to Chuang-zi, Lao-zi, and various other thinkers in the tradition of Chinese Taoism; in her 1973 article "Thoreau: An American Taoist Sage," Patricia A. Okker, for instance, points to social criticism, philosophy of nature in *Walden*, and belief in a kind of wu wei (the nonaction implied through Thoreau's argument in favor of simplicity) as ideas that parallel Lao-zi's *Tao Te Ching* (86, 92). Other scholars have traced a wide range of intersections between East Asian philosophy and Western environmental literary expression. For example, in his 2013 article "Original Nature: Buddhism and American Nature Writing," David Landis Barnhill writes in depth about two of Lopez's peers, Gary Snyder and Peter Matthiessen, who have been powerfully invested in Buddhist ideas and practice and have woven Buddhist perspectives throughout their literary work. Barnhill argues that Buddhism has offered Snyder and Matthiessen "a distinctive cultural lens for understanding the natural world, our relationship to it, the social context of our lives, and our responsibilities to the Great Earth Sangha, the sacred community of life," a lens that is, in different ways for each writer, personal, mystical, and ecological (94).

The point I would particularly like to make, however, is that "new ecological perspectives from East Asia" may not always be "new" and may not even seem, on the surface, to be "ecological."

Think of the enhanced sense of reverence toward the nonhuman that appears so vividly in the work of Barry Lopez and many other American writers who have directly or indirectly received inspiration from Eastern thought, from Kenneth Rexroth and Jack Kerouac to Jane Hirshfield and Red Pine (Bill Porter). Such heightened respect is not a new idea at all but may have drifted through history from a variety of sources, including various Confucian channels. Think, too, of how Lopez attaches this reverential perspective not only to the natural world but to language itself, to a broad array of subjects (including technology), and even to his readers. The newness of an idea adopted from another culture, near or far, depends on the freshness of a writer's own vision and articulation, not on the actual originality of the adopted concept. Likewise, an "ecological perspective," for many thinkers, depends less on an understanding of the natural world beyond the human realm and more on a fundamental sense of interconnectedness. In recent years, North American environmental thought has become more and more devoted to this notion of inextricable connection, the sense that our bodies incessantly give to and take from the body of the world and even the sense that agency and narrative meaning are inherent not only to human consciousness but to the wider animate and physical world.

As I have suggested above, there is a proud tradition in North American environmental thought and expression of adopting fundamental ideas from East Asia (and from other cultures throughout the world). This syncretic merging of home-grown experience and exotic inspiration has been happening since the very beginning of American environmentalism. However, Xinmin Liu and Peter I-min Huang have now compiled a fascinating collection that exhibits a much broader range of East Asian ecological thought than has previously been accessible to readers outside of Asia, including different ways of understanding our relationship to nonhuman animals and alternative concepts of landscape. I believe exposure to the exciting ideas presented here will super-charge the absorption of East Asian perspectives in ecological thinking elsewhere in the world.

Eugene, Oregon

REFERENCES

Barnhill, David Landis. "Original Nature: Buddhism and American Nature Writing." In *Critical Insights: Nature and the Environment*, edited by Scott Slovic. Salem Press, 2013, pp. 78–99.

Keefer, Bob. "Lopez's Book on Arctic is Best in America." *The Eugene Register-Guard*, 19 November 1986, 1B.

Lopez, Barry. *Arctic Dreams: Imagination and Desire in a Northern Landscape.* CharlesScribner's Sons, 1986.

O'Connell, Nicholas. "Barry Lopez." In *At the Field's End: Interviews with 20 Pacific Northwest Writers*. Madrona, 1987, pp. 3–18.

Okker, Patricia A. "Thoreau: An American Taoist Sage." *The Comparatist*, May 11, 1987, pp. 86–95.

Paul, Sherman. "Making the Turn: Rereading Barry Lopez." In *Hewing to Experience: Essays and Reviews on Recent American Poetry and Poetics, Nature and Culture*. University of Iowa Press, 1989, pp. 343–386.

Acknowledgments

To begin with, we would like to thank Kathryn Yalan Chang, Kiu-wai Chu, Simon Estok, Peter I-min Huang, Xian Huang, Young-Hyun Lee, Hua Li, Jialuan Li, Xinmin Liu, Xin Ning, Kenichi Noda, Iris Ralph, Scott Slovic, Lili Song, Ban Wang, David Wang, Qingqi Wei, Philip F. Williams, and Sijia Yao for their inspiring and innovative contributions.

We would also like to express our deepest gratitude to Scott Slovic for writing the Foreword and to Xiao-Hua Wang, Chia-ju Chang, Yuki Masami, and DJ Lee for the endorsement of our book.

Special thanks must go to Douglas Vakoch, series editor of Environment and Society of Lexington Books, for including this book in their environmental studies series—Asian Studies. We are also indebted to the generous and efficient assistance of the editorial, production, and marketing teams of Lexington, especially to Kasey Beduhn, acquisitions editor, and Vishnu Prasad R, and Dominique McIndoe for overseeing all production process to the published volume.

Hanqian Xu also deserves a note of thanks for assisting the co-editors in managing the editorial archive in the early days of this book project.

This book is "the greatest milestone to mark our collaboration and friendship since the summer of 2012," as Xinmin recently wrote to Peter in an email. It is almost a decade since we first met at a conference we attended at Shandong University, China. As we trace its evolving path in retrospect, it gives us a memorable occasion to express our indebtedness to the many academic institutions on both shores of the Pacific that kindly provided funding to boost the growth and maturing of the key ideas and notions underpinning this volume. We thank profusely Office of Research Advancement at the Washington State University, Grant and Fellowship Support Program of College of Arts and Sciences at WSU, Center for Chinese Studies at the National Central Library of ROC, Taiwan, and Annenburg School for Communication at the University of Pennsylvania. The book came to its early shape at "A Symposium on S-F/Sci-Fi and Planetary Healing" at Tamkang University in the winter of 2016. We are grateful to many distinguished scholars from East Asia and the United States who joined our endeavor, enriching our ideas with theirs and nurturing our collective wisdom from seedling to fruition along the way. Finally, our thanks go to our families and mentors, and activists and the environmental community at large. They continually inspire us in such calls as these to heal the planet.

Introduction*

Xinmin Liu and Peter I-min Huang

Within a brief span of about three decades, China's industrial development and economic growth moved in leaps and bounds, earning her the much coveted status of one of the world's leading economic powers in a century. Yet it is difficult to believe that the break-neck pace at which China leapfrogged through a growth boom driven by the global market has been brought about without a host of menacing aftermaths, that is, interfering and harmful impacts, on her people, their living habitats and, particularly, on the centuries-old human–land ties and the heritage of cosmological wisdom. Attitudes and mindsets toward such ecological costs mark a watershed moment, the moment at which we must question ourselves how we came to land ourselves in this ecological mayhem?[1] What must we do right now to let the wellbeing of the human and nonhuman lifeworld bounce back? There is indeed no time to be lost in challenging ourselves to own up to the severity of the environmental degradation, defog our blurred vision by ridding ourselves of many of the conceptual blinders and dissonances, and hopefully we will stay the course in preserving and recovering our shared livelihoods on this planet in a responsible, healthful, and sustainable way.

A few words about where we are coming from; in the spring of 2017, as a group of motivated environmental humanists, we reached across geographical distances and disciplinary boundaries to join hands in pursuit of a shared goal: to publish a "Special Issue of *SF* (speculative fiction) and Planetary Healing" curtesy of *Tamkang Review* in Taiwan. In the "Foreword," we stated:

> [W]e do not avert the climatological and geological *agency* of the humans, but will posit it as a key venue of seeking to heal and recover our planetary existence. We will explore how to value planetary limitations as ethical categories

---

* Beginning with Page 8 to Page 12, the text is written by Peter I-min Huang.

through historical memories and cultural heritages by virtue of reflective nostalgia; we will affirm rather than renege an embedded mode of life so that our
sense of affect and empathy is always nurtured in concert with the vitality of
our earthly belongings.[2]

In retrospect, we are much relieved to notice that awareness has been raised
significantly as we humans have come to reckon with the fact that we are
guilty as charged with the geological and climatological damage and ruin,
and that, in the same vein, we need to remain vigilant against the lingering
denial of and resistance to planetary limitations. We must stay active and
assertive in striving for a responsible mode of living and enhancing our moral
sensibility and ethical obligation and in contributing to a more robust drive of
ecological healing across Asia and the rest of the world.

We also believe that healing has not thus been made *any less* daunting,
pressing, and arduous. Presently, as we speak, we are having a foretaste of
what we will be running against in the near future, that is, facing adversi-
ties such as distraction, indifference, and even bias against social advocacy
of ecological healing—for which we are not out of the woods yet. Over
the past year or so, no matter our geographical locations, health care provi-
sions, and physical and mental wellbeing, we have all fallen victim to the
outbreak of COVID-19 pandemic, which has not only placed our normalcy
of working, traveling, and living under life-altering risks and dangers across
the globe, but exposed us to wave after wave of assaults owing to social
inequality, racial and gender biases, science-phobia, xenophobia, renewed
cold-war ideologies, and meltdowns in societal infrastructures and political
governance.

There is, nonetheless, a silver lining out of this viral pandemic: In most
ironical terms, the zoonotic jump of COVID virus at its origin, the likelihood
of its across-species spread from some wild animals to the global human
population has at last brought the public so much closer to the stark reality
that humans are ineluctably part of the animal world, and that the borders
between the humans with the nonhuman lifeworld are porous and vicarious—
a sobering reminder that, in corporeal terms, we humans have not put much
evolutionary distance between us and the rest of the earthly lifeworld. Thus,
we need to ask: what is it that has caused humanity to lose sight of such a
profound yet vital piece of the evolutionary puzzle? What risks, pitfalls, and
perils have we tolerated, indulged, or even ignored in our pursuit for civiliza-
tional progress that has led us to such blinding and costly misconception of
the human–nonhuman ties of coexistence? And in what ways can we begin
to remedy the mishaps and right the wrongs before the tipping point of the
biophysical balance ignited by humans' lopsided, runaway despoliation of
the ecosystem evolved originally to sustain all living creatures on the Earth?

Restoring the impaired human visceral bond to the natural world is the key. And we believe that we cannot be effectively reconnected to our biophysical world unless we go about bonding with it by way of an embodied and embedded mode of living. To do so, we need first and foremost to straighten out our warped understanding as to how we humans are interconnected to the natural world. The Chinese scholar Fan Meijun remarks: "Nature is a lovable object for appreciation." Fan takes it for granted that nature is not only to be considered as an object for appreciation, but the act of appreciating nature as an external object is "an indispensable part of man's spiritual life" (267). There is clearly a presumed intent to posit a binary on Fan's part while configuring the human–nature correlation, and it follows that the human act of appreciating the values of nature is conduced as one based on the cognitive divide between the subject vs. object and materiality vs. spirituality. Fan goes further with this binary by asserting that in ancient China theories are based on *instinct* whereas in the West theories are based on *science* (269). It is no surprise that, when comparing the ecological consciousness in ancient China with that in the modern West, Fan has this to say:

> Third, the contents of their theories are not the same. The ancient Chinese ecological consciousness is related to appreciation of the beautiful; therefore it is concerned with people's spiritual life. It stresses the point that nature offers a spiritual reward to man by purifying and upgrading man's soul for his esteem for and appreciation of nature. In contrast to this, the Western ecological consciousness is mainly concerned about man's existence. It emphasizes that nature will punish human beings if they keep plundering it. It shows a special concern about how human beings will be able to survive natural disasters. (270)

A cursory look at the author's point of view as conveyed above convinces us that Fan has dubiously implied that *science* (of the West) has outgrown over *instinct* (of ancient China), and that Fan has therefore aligned the two modes of historical growth with a single-track, linear path of historical development, namely, a law of irreversible progress enabling the scientific modern West to outlast the instinct-driven ancient China. Thus, Fan has joined with many others prior not only in favoring industrialization over agriculture, rationality over spontaneity, but more importantly, in endorsing the West as a teleological destiny for orienting China and other developing nations the world over toward an ultimate realization of West-led modernity.

Fan's suggestion of historical inevitability in defining the human world and the nonhuman nature is evidently not adequate because it attempts to transplant in retrospect the notion of "aesthetic appreciation" in the context of ancient China, and the transplanted notion was enmeshed in such a way as if aesthetic appreciation had been the unmatched primary aspect of cultural values in humans' dealing with the nonhuman nature. Nothing could be further from the truth; as Tu Wei-ming has contended, "the Chinese metaphysical assumption is significantly different from the Cartesian dichotomy between spirit and matter" (68). A renowned scholar of early Chinese thought, Tu has famously critiqued key historical and cultural theories advanced by famed sinologists, such as Joseph Needham, and F. W. Mote, and he has taken a firm stand toward the Confucian emphasis on *ethical righteousness* as the foremost distinction of ancient Chinese thought. While theorizing his notion of "Continuity of Being," he states:

> Humanity is the respectful son or daughter of the cosmic process. This humanistic vision is distinctively Confucian in character. It contrasts sharply with the Taoist idea of non-interference on the one hand and the Buddhist concept of detachment on the other. Yet the notion of humanity as forming one body with the universe has been so widely accepted by the Chinese, in popular as well as elite culture, that it can very well be characterized as a general Chinese world view (74).

It is perfectly clear that Tu unequivocally regards the nature of the human–nonhuman correlation as chiefly ethical and relational, and that the role of humans in interacting with the nonhuman nature is decidedly participatory as defined by the part-to-whole continuum. It thus goes to prove Fan's notion inadequate and misleading in setting the viewer/reader apart, mentally and physically, from the nonhuman natural world and limiting the human interaction with nature to a disembodied act of "appreciating" beautiful landscape.

This brings us to the following questions: How does the appreciation of landscape beauty lead to a reverence of the natural world as a whole? And how can beautiful landscapes give us more life-giving and life-sustaining power than the eye-pleasing relief and gratification they momentarily bring? We humans are known to have loved the "beauty of nature" for as long as we've owned written records of history, but in recent research findings on the environmental history of China, scholars have revealed an "ingrained" tendency to peruse the mountains and waters (*Shanshui* 山水) in certain *pictorialized* and *mentalized* manner that is likely to estrange the viewer/reader from the landscape itself and appreciate it as mere mind-soothing scenic beauty, especially when it is digitally enhanced with CGI, that one needs occasionally and leisurely for their stress relief and their escape from the day-to-day humdrum. This has prompted many scholars to dub the tendency as "ecophobia" (恐惧生态症)—an aesthetic mode that retiringly stimulates, but often miscues in the viewer illusions of loving scenic beauty via eyesight or automatic viewing device *alone* while the viewer remains sedately withdrawn from any visceral contact with the lifeworld (Estok 2011; Elvin 2004). In other words, only eyesight, and no other human sensory organs, are involved in this nature-appreciating act and, as long as the human mind elevates only the visually captured image to an abstract or aesthetic level, the human viewer does not even have to take a single step outdoor to be physically present at the site of the actual scenery.

To us, what lies front and center in this *ecophobic* mindset is the humans' penchant to privilege our eyesight *over* other sense organs when configuring the biophysical world. This cognitive privilege is sometimes savored to such an indulgent extent that it misguides or even interferes with the rest of our senses, corporeal organs and the brain—as is proven true with the birth of digitally simulated virtual reality (VR)! This is precisely the artistic mode that retiringly stimulates, but often miscues illusions of "loving the beauty of nature" by means of eyesight alone and being sedately withdrawn from bodily contact with the nonhuman nature. The culmination of such a topophobic trend in China's dynastic art heritage is found in one painting by the late-Ming dramatist Li Yu 李渔 (1611–1680) who wrote in his *A Temporary Lodge for My Leisure Thoughts* (*Xian qing ou ji* 閒情偶记) that he had invented a "landscape window," a replicated image of his most favored outdoor landscape, which he painted and installed to cover up the original window of his study. Not surprisingly, the design of Li's original painted scene of *Shanshui* was before long canonized and compiled under the name of *Album of the Garden of Mustard* (*Jie zi yuan ji* 芥子园集) and have since been studied, traced and emulated by millions of Chinese. There is little doubt where Li Yu's concept of "Windowed Shanshui" was headed for, and we moderns, when visually wired to the extreme, have been turned into particularly "insensitive" and vulnerable offspring of Li. Since Western-led modern values set foot in China around the 1840s, we have carried the trend of pictorializing natural scenery to an extreme, for this eye-mind coalition is apt to short-circuit our already vision-centered cognitive process, and without its being plugged in to hearing, smelling, touching, and tasting, the human grasp of the nonhuman surrounding is partial, lopsided, and ultimately downgraded to a total disconnect!

This has made it necessary to revisit our earlier critique in order to disclose how a sight–mind short circuiting of nature appreciation had transpired in Fan's theorizing of nature. And it is through our reading of Tu's assertion of Continuity of Being that we have discovered the truth in clarity: as China's early modernists embraced the Cartesian perspectivism *in toto*, we have worn down our alertness to our internalized preference for the subject–object binary; moreover, we have neglected the proper function of the human episteme as a blending of all sensory reactions, which requires our intellect embody an assortment of sensory interactions with our inhabited biophysical world.

To be clear, Tu's advocacy of Continuity of Being does embrace the role of seeking spiritual gratification, and it is therefore in no position to preclude the human need to transcend one's corporeal limits in search of things beautiful in mental as well as in physical states of being. But the soaring flight of transcendental imagination should never be lifted off humans' visceral and terrestrial moorings once and for all. As he believes, "That humankind receives *Ch'i* in its highest excellence is not only manifested in intelligence but also in sensitivity. The idea that humans are the most sentient beings in the universe features prominently in Chinese thought" (75). To transform the sensitivity into lived and embodied praxis, he states, it is by means of "affect and response" (*kan-ying* 感应), whose function he elaborates as follows:

> The mind forms a union with nature by extending itself *metonymically*. Its aesthetic appreciation of nature is neither an appreciation of the object by the subject nor an imposition of the subject on the object, but the merging of the self into an expanded reality through transformation and participation. (77, authors' emphasis)

Tu's stress on the human affective shared with all sentient creatures is precisely what enables us to further dissect how appreciating natural beauty by way of qi (ch'i 气) entails that humans rely centrally on our sensory organs, bodily parts and the corporeal being as a whole to stay in tune with the beauty and vitality of nature. Defined by Cheng Hao chiefly as the "vital force" that circulates through all forms of life on the planet, qi is what blends and melds humans and the natural world into one unified entity; it follows that "human life is part of a continuous flow of the blood and breath that constitutes the cosmic process. Human beings are thus organically connected with rocks, trees, and animals."[3] Taking que from this notion of qi, we now proceed to read the idea of "Qi-transformed" shanshui by the Taiwan Confucian scholar Yang Ru-bin (楊儒賓 1956–). Yang has devoted his scholarly career to researching on Confucian ideas about the human–land correlation and has played a ground-breaking role in enhancing the impact of qi. Among his many published works, two edited volumes stand out: *Confucian Theories on Ch'i and Wellness Cultivation* 儒學的氣論與功夫論（2012）and *On the Evolving Concepts of Nature* 自然概念史論 (2014). Both books deal with qi-related concepts of Confucian principles of philosophical, ethical and cultural significance with particular focus on the *embodied* and *experiential* poetic imaginary of many landscape-inspired poets and writers, past and present. But his astonishing breakthrough came earlier in an essay published in 2009 in the *National Taiwan University Academic Journal in Chinese*, to which Yang gave the title "How was Shan-Shui Found?: The Analysis of "Xuanized Shanshui ("山水是怎麼發現的": 玄化山水析論).[4] With a friskily posed question to begin the essay, Yang loses no time in declaring that his question is not meant to reset the discussion as to when and how Shanshui had first been discovered in China's dynastic literary tradition; rather, he is reclaiming the grounds for a more embodied and embedded mindset as he traces its trajectory in the works of his favorable shanshui poets. He chooses to call this rediscovered poetic imaginary "xuan-ized Shanshui" (玄化山水), but later avoids using the term in favor of "Qi-transformed shanshui" (气化山水), which, as his self-coined neologism, resonates far better with his other qi-themed terms such like "Qi gan" 气感 (qi-induced feelings) and "Qi xiang" 气象 (qi-permeated milieu) used in the latter half of his essay. Not to be confused with "Xuan" (玄), which denotes a metaphysically alleviated state of mind, his shifting to the "qi-transformed" mode rings off a robust and re-energized voice, and conjoins it with the powerful chorus by ecocritics who for the past decade or so have actively contributed to the discussion of "atmospheric thinking."[5] It is therefore quite fitting for us to take a closer look at the steps Yang lays out so as to accomplish a qi-driven transformation for humans when conceptualizing the natural world.

Specifically, to shed fresh light on the function of shanshui, Yang tries to energize anew the organismic flow of force across the human–nonhuman rift by means of a qi ambience engulfing all residing in the natural world, yet he refocuses the gravity of configuring nature on the poet as a sentient being "enacting it bodily" in it rather than a cool onlooker observing from above and beyond it. He takes the lofty-minded poets to school by revealing to them that "only those who let their minds be infused in qi-induced organismic force are capable of savoring nature's animistic worth" 只有存着虚化之心眼者，才可以领略山川之美 (Yang 235). But the question remains as to how the embedded observer begins his/her initial act to be transformed by the qi ambience? And in what ways does this act induce the observer to embrace a partaking of the natural world with a sense of belonging? To that end, let's sample a few key tasks from his must-do list; we find first and foremost that Yang urges the poets to "[B]e qi-consumed so as to find joy in the qi-saturated world" 含灵以趣灵, and to bring alive their corporeal being to open and expand themselves bodily and mentally 散怀, and that to let their affective capacity expand totally and gratify to the full 畅情, and to glance freely and widely 游目. It is no surprise that these words (especially the action verbs in bold fonts) intend to mobilize the humans' bodily enactment, which leads further to a merging of the viewer and the viewed. Yang's call reminds us of Yi-fu Tuan, the cultural geographer, who appeals to us to put into practice "synesthesia" by activating all our sensory functions concurrently when conceptualizing our natural terrains (Tuan 366). Sharing a similar mindset with Tuan, Yang writes: "phrases like 散怀 and 畅情 read more like a compounded sentiment as a result of mingling various bodily reactions of one's mind and body. It can be regarded as the feeling of one's entire corporeal existence" (236). To be expected, both scholars have placed premium on the human bodily enactment as the main premise of nature exploration by the humans. Apparently, these vigorous enactments of bodily postures and mental exercise, it might be disputed, are not uncommon to certain extraordinary landscape poetry and paintings. What is different with them are as follows: (a) they signal the beginning of a participatory involvement of all the observer's senses concurrently and would lead to an interactive configuration of shanshui between the humans and the real-life nature; (b) they would rarely end up stranded uprooted and lifeless at the transcended height and never to come back down to earth, because these actions are a process of robustly renewing and infinitely becoming. Despite geographical distance and age difference, Tu and Yang have made admirable breakthroughs in their shared realm of conceptual affect, thus shedding fresh light on the diverse and animated minds of scholars, we feel so honored to gather in the ensuing chapters of this volume. It is gratifying to say that we are inspired and enriched by their views and approaches that form a rich spectrum of dynamic and insightful reflections of these two pathbreakers.

In opening the volume with six chapters grouped under the subheading, "Living Wisdom and Lived Heritages," we foreground ancient East Asian cultural beliefs and practices that are inspiring and shaping the environmental consciousness in the twenty-first century and so contributing to the "resacralization of the cosmos," to borrow ecocritic Kate Rigby's wonderful words (287).[6] The six chapters are complete in themselves; they also establish the theme of the entire collection, emphasizing that the challenge to modern Euro-Western onto-epistemologies—under the aegis which the nonhuman world is regarded essentially to be an object without purpose and redeemable only when commodified—is critically funded as much by ancient as by (post-)modern environmental practices. In chapter 1 that inaugurates this anthology, David Wang traces the origins of the present environmental crisis to the Baconian instrumentalist view of nature that dominated Western thought from the sixteenth century onward and then looks at earlier diametrically opposite beliefs. These are found in the teachings of the Neo-Confucian philosopher, Zhu Xi (960–1279 CE), and Christian apostle, Saint Paul (ca. 5–64 CE); they emphasize that the human is a mere component, and indeed a small "participant" in the natural "enormity" of things and what is most viable, or sustainable, is modesty, humility, and proportion. In chapter 2, Ban Wang provides a fascinating and provocative discussion of Lu Xun (1881–1936), a figure who played a significant role in China's anti-imperialist May Fourth movement. A "prescient critic of the Enlightenment, science and technology," what Lu Xun lambasted was not scientific and technological knowledge per se but a technocratic elite and gentrified class of people who, like their bourgeois counterparts in the West, touted progress and the democratization of private wealth and power but in fact sought more to access that for themselves through new markets of science and technology. Lu Xun's call for the revival and restoration of older cultural practices and beliefs—inclusive of those richly preserved in songs, poems, and other artistic expression—reflects, as Wang persuasively argues, an appreciation of human relationships with the natural world built on humans' respect for nature's own unique residencies of power, will, and agency. Chapter 3 emphasizes that premodern beliefs—namely those taught under Daoism—offer the modern-day environmental movement unique insights into how to restore more balanced relations between human and nonhuman existence. Describing Daoism in the context of its coincidental emergence with the Agricultural Revolution of "some 8,000 to 10,000 years ago," an era in human history that Li and Wei distinguish according to the term "the minor Anthropocene," the authors discuss how Daoist tenets reflect contemporary environmental efforts to achieve an "ideal state of equilibrium" and particularly so in the face of the present era of the "severe Anthropocene" (which scholars generally agree to be an epoch that either begins with the Industrial Revolution or emerges in the post-World War II era and ties to, among other anthropogenic phenomena, nuclear testing in 1945 and the detonation of a nuclear bombs).

The three chapters that make up the first half of part I address Confucianism and Daoism, two towering pillars of Chinese culture and society. In chapter 4, Lili Song calls attention to a less well-known anchor of East Asian culture and society—namely, ancient animist beliefs. Disputing the common bias that animism, as it appears and distinguishes minority literatures, betrays spurious, backward looking, and superstitious beliefs, Song, in a reading of two works of minority literature by a Mongolian and Ewenki author respectively, Guo Xuebo and Chi Zijian, gives center stage to what literary and cultural studies scholars typically place backstage or in the wings, namely, notice of the vital role that animism in minority literature can and does play in the work of environmental healing.

A particularly germane genre in the twenty-first century, science fiction (sci-fi) is the inspiration of the final two chapters of part I. In chapter 5, Hua Li compellingly argues for a shift in mainstream attitudes toward science and technology, from stances that betray profound ignorance or dismissal of the ethical and other moral costs of more or less regarding all nonhuman life as being a petri dish and potentially consumable product to positions that underscore the inherent value of all living beings. As part of this argument, Li singles out two contemporary works of sci-fi representing the shift in some Chinese sci-fi since the 1990s. Between the 1950s and 1980s, techno-national belief in the ability of modern science and technology to solve issues of food scarcity and related concerns was at an all-time high. Sci-fi in the last thirty years continues to tout the capacities of human made inventions but also reflects work, such as the novels of He Xi, that denote a more cautious approach to and even skepticism of the utopic belief that modern science and technology can achieve what no other forms of human knowledge let alone other-than-human knowledge can accomplish. Chapter 6 by Peter I-min Huang, the last chapter in part I, is a compelling critique of urban-industrial policies and practices and efforts by environmentalists, inclusive of poets and other writers and artists, to preserve and rehabilitate green space. In addition to canvassing poems by Guangzhong Yu, Ming Xiang, Qiao Zhong, and Hung Hung, Huang reads international best seller sci-fi and cli-fi (climate fiction) novel, Ming-yi Wu's *The Man with the Compound Eyes*. Similar to the minority literature that Lili Song writes about, Wu's novel, as Huang highlights, calls attention to and celebrates ancient indigenous forms of environmental praxis and how that vitally supports rather than slows down or stands in the way of moving forward as a species.

The five chapters that comprise part II of the collection, entitled "The Embodied Imaginary," speak for new directions in environmental thinking, trajectories that owe to (post-)modern realities inclusive of advances in the relatively new area of biotechnology and to the all but omnipresent phenomenon of the visual medium. The first two chapters cast critical light on the dark synergies between genome research, production of transgenic matter, and the global food industry. The objects of cynosure of the last three chapters are cinematic and other visual texts that critically overturn or tamper with notions of so-called objective and universal perspective. Chapter 7, by Simon Estok and Lee Young-Hyun, is a dazzling *tour-de-force* critique of a novel that, while it has received widespread critical attention, mostly has been praised in lavish and unstinting terms. Estok and Lee, in focusing their reading on the main character of Emiko, an abandoned bioengineered cyborg invention and creature, draw attention to how the novel betrays and implicitly caters to "Western xenophobia" and "Western fantasies and stereotypes of Asia" as well as to how the novel capitalizes on current debates about genetic manipulation but hardly engages with the environmental issue of the staggering loss of gene pools, namely those that distinguish biodiversity. Iris Ralph's The BBC Drama Series *ShakespeaRe-Told* and Eric Yoshiaki Dando's *Oink, Oink, Oink*, brings together ecocriticism, (critical) animal studies, and biosemiotics theory. Focusing on the porcine species (one of the most industrialized and experimented animals on the planet), carnophallogocentricism (a term coined by Jacques Derrida), and the issue of humans' denial of language to the nonhuman world on the basis of anthropocentric definitions of language, Ralph contrasts rampant speciesism in the present century with environmental calls for and practices of multi-speciesism. The third chapter in part II, chapter 9 by Kenichi Noda, enlists the term "logic of glance," in an argument about what "multiple points of view," or nonperspectival modes of seeing and "unmediated" forms of "visual perception," disclose in comparison with what linear perspectives dictate. Challenging the "scopic regime" of linear perspective in the period of modernity, Noda illustrates his powerful arguments in a discussion of alternative, ecocritical, perspectives as these appear in representations of the landscape in such novels as Ooka Shohei's *Wildfires.* Sijia Yao, in chapter 10, analyzes the aerial gaze as this distinguishes several documentary films: *Beyond Beauty* (2013), *China from Above* (2015), and *Aerial China* (2017). As Yao cogently argues, documentary films are paradoxically ciphers and decipders of truth, for governments can use them meretriciously, mostly promoting nationalistic agendas and human exceptionalism instead of educating the public about such issues as environmental degradation. Chapter 11, the final chapter in part II, by Kiu-wai Chu, also explores the ways in which film—in this case, a commercial film—both serves and betrays environmental projects and aims. Chu's main subjects of inquiry include the Tanka boat people of Hong Kong as they are represented in film director Stephen Chow's blockbuster, *The Mermaid,* and the attrition of the pink dolphins in Hong Kong (in the waters near Tai-O and some offshore islands and in the Pearl River Estuary). Tracing Hong Kong identity to the ancient legend of the Lo Ting, half-human and half-fish hermaphrodite beings, Chu navigates the murkier channels of the glamorous world of global capitalist film production. Many films that identify with this production promote and draw attention to pressing environmental issues, as in the case of *The Mermaid*, which reflects particular awareness of environmentalists who are campaigning for a multispecies ethics and for environmental forms of empathy ("eco-empathy"). However, in capitalizing on those issues many films that carry an environmental message do not engage further with those issue, or in ways that evidence sincere, sustained, and educated commitment to restoring, reviving, and healing the green world.

In Chinese, there is a saying, "The best wine is at the bottom of the jug." And so this anthology would not be replete without the four chapters that comprise the third and final section. Entitled "Myriad Therapeutic Lands," it brings together the issues of ancient environmental wisdom and modern environmental thinking that part I and part II, respectively, showcase. In chapter 12, Kathryn Yalan Chang discusses eco-displacement in the context of the nuclear power plant accident in Japan that devastated this East Asian country. Bringing to mind the work of ecocritic Serenella Iovino and the concept of "storied matter," Chang contrasts the "apocalyptic logic" of the Anthropocene inclusive of reliance on nuclear power plants with the notion of the "therapeutic landscape," or places that provide affective and material support as well as powers of healing. In contrast with, and in implicit opposition to, Chang's chapter, chapter 13, the most controversial chapter in this volume, constitutes a vigorous defense of nuclear power and of how it necessarily complements renewable sources of power inclusive of solar, wind, and hydro power. Some or all of the other contributors to this volume might take issue with Williams's claims yet also recognize the need for openness to discussion of a source of power saddled with a bad reputation among environmentalists (for among other reasons, the unresolved problem of nuclear waste disposal). In his advocacy for the use of nuclear power, Williams also comments on several works of Chinese sci-fi that include the novel, *Ball Lightning*, by Liu Cixin, the acclaimed author of another sci-fi, *The Three-Body Problem.* Williams applauds Liu's evident distaste for or at least suspicion of environmental activists, who are represented in both of these novels as being deluded luddites and ticking-time-bomb terrorists.[7]

In chapter 14, Ning Xin addresses the term and concept of urban ecology in the context of "city resurrection" projects in Beijing, China. Eliminating the old non-Cartesian city spaces with their labyrinthine narrow streets, and replacing them with expensive apartments and spectacular skyscrapers, the builders of the new "Chinese BoBo—Bourgeoisie and Bohemian" city spaces reflect neo-liberalist policies that may promise more than what they can deliver, as Ning explains. Chapter 15, by Xian Huang, concludes part III. Similar to Ning, Huang addresses urban ecology initiatives and projects in China. Differently from Ning, Huang focuses on issues of geomedia and new forms of locality that are both embodied and disembodied, giving people a sense of place in digital more than material terms. In representing the final contribution, Huang's chapter responds to one of the major challenges of the present century, how to hold on, preserve, and heal old material worlds in East Asia and elsewhere on the planet in the face of seemingly overwhelming calls to rampantly dismantle those worlds and erect virtual realities in their stead.

NOTES

1. There is indeed no time to be lost in challenging ourselves to own up to the severity of the environmental degradation, defog our blurred vision by cleansing many of conceptual blinders and dissonances, and hopefully we will stay the course in preserving and recovering our shared livelihoods on this planet in a responsible, healthful and sustainable way.

2. Xinmin Liu, "Foreword," *Tamkang Review: A Journal of Literary and Cultural Studies* (Taipei, TW: Tamkang University) Vol. 47, No. 2, June 2017, iv.

3. Cheng Hao 程灏 （1032–1085）. This quotation from Cheng Hao is cited by Wei-ming Tu in his essay "The Continuity of Being" collected in *Nature in Asian*

Traditions of Thought: Essays in Environmental Philosophy (Albany, NY: State University of New York, 1989), 74.

4. Ru-bin Yang, "How Was 'Shan-Shui' Found?: The Analysis of 'Xuan-ized Shan-Shui'," (Taipei, Taiwan: National Taiwan University) No. 30, June, 2009, 209–254. Doi: 10.6281/NTUCL.2009.30.06.

5. Of the many such scholars, I reference two here: Jesse Oak Taylor, author of *The Sky of Our Manufacture: The London Fog in British Fiction from Dickens to Woolf* (Charlottesville: University of Virginia Press, 2015), and Katherine Anderson, *Predicting the Weather: Victorians and the Science of Meteorology* (Chicago: University of Chicago Press, 2005).

6. Kate Rigby, "Spirits that Matter," in *Material Ecocriticism*, ed. Serenella Iovino and Serpil Oppermann (Bloomington: Indiana University Press, 2014), 283–290.

7. For a different reading of *The Three-Body Problem*, one that argues in effect that Liu disingenuously misrepresents environmental activists, see Peter I-min Huang's "Chinese Science Fiction and Representations of Ecofeminists: Mad Women or Women Warriors," in *Ecofeminist Science Fiction: International Perspectives on Gender, Ecology, and Literature*, ed. Douglas A. Vakoch (London: Routledge, 2021).

WORKS CITED

Anderson, Katherine. *Predicting the Weather:Victorians and the Science of Meteorology*. Chicago: University of Chicago Press, 2005.

Huang, Peter I-min. "Chinese Science Fiction and Representations of Ecofeminist: Madwomen or Women Warriors." In *Ecofeminist Science Fiction: International Perspectives on Gender*, ed. Douglas A. Vakoch. London: Routledge, 2021, 127–138.

Liu, Xinmin. "Foreword." *Tamkang Review: A Journal of Literary and Cultural Studies* (Taipei, TW: Tamkang University) Vol. 47, No. 2, June 2017, iv.

Rigby, Kate. "Spirits that Matter." In *Material Ecocriticism*, ed. Serenella Iovino and Serpil Oppermann, 283–290. Bloomington: Indiana University Press, 2014.

Taylor, Jesse Oak. *The Sky of Our Manufacture: The London Fog in British Fiction from Dickens to Woolf*. Charlottesville: University of Virginia Press, 2015.

Tu, Wei-ming. "The Continuity of Being." In *Nature in Asian Traditions of Thought: Essays in Environmental Philosophy*, ed. J. Baird Callicott and Roger T. Ames, 74. Albany, NY: State University of New York, 1989.

Yang, Ru-bin. "How Was 'Shan-Shui' Found?: The Analysis of 'Xuan-ized Shan-Shui." *National Taiwan University Academic Journal in Chinese* (Taipei, TW: National Taiwan University) No. 30, June 2009, 209–254. Doi:10.6281/ NTUCL.2009.30.06.

Part I

LIVING WISDOM &
LIVED HERITAGES

Chapter 1

Humility by Proportion

What Zhu Xi and St. Paul Have to Say about the Baconian Attack on "Nature"

David Wang

This chapter critiques the Baconian view of "nature," suggesting the damage it has done to the broad range of theory and praxis subsumed under the banner of sustainability in general, and respect for ecological processes in particular. As to what I mean by ecology and sustainability, suffice it to say that I appreciate Aldo Leopold's notion of *a land ethic*. Leopold envisions both natural processes and human being[1] in a single biotic unity, one that is organically self-sustaining and, if left alone, thrives abundantly in growth.[2] The entirety of this organic system Leopold calls the *land pyramid*,[3] and living proportionally within the distributions of this land pyramid constitutes the land ethic. It is an ecological ethic, and sustainability comprises those values and actions that maintain the health of this organic holism.

The opposite of "if left alone" is, in general, the intrusions of technology and proliferation of capital that industrial and postindustrial manipulations of nature for human comfort have wrought. But here, I do not promote cessation of these factors because it would be unrealistic at not only a global scale, but even at local scales. The "elephant in the room" in much of the literature calling for a return to union between our bodies and the natural cosmos is that even we who write about these things do not realize just how separated we are from the nature around us. I have in mind Charles Taylor's diagnosis of "the secular age."[4] While we often divide between those who are religious from those who are secular, Taylor argues that we are *all* secular inasmuch as we all live in what he calls an *immanent frame*.[5] This is a condition in which the cosmos is perceived as purely material, to be manipulated for human comfort, with no transcendent presence in it at all. Those who are religious might have an "open construal" to the idea that there just might be transcendence. Those with a "closed construal" reject even the possibility of anything beyond the immanent web of socially constructed meanings.[6] For Tayler, academics are prone to favor this closed view.[7] But for all in contemporary life, manipulating nature for human comfort deeply informs social praxes.

In contrast to this received view, I suggest it is difficult to consider matters of ecology and sustainability without recognizing there is something absolutely mysterious inherent in the natural world. Here is the conservationist Wendell Berry writing on a basic component of the natural world, topsoil:

> It is apparently impossible to make an adequate description of topsoil in the sort of language that we have come to call "scientific." For, although any soil sample can be reduced to its inert quantities, a handful of the real thing has life in it; it is full of living creatures. And if we try to describe the behavior of that life we will see that it is doing something that, if we are not careful, we will call "unearthly": *it is making life out of death.* Not so very long ago . . . we would probably have called it "miraculous."[8]

In light of this, rather than unrealistic global change in matters of ecological sensitivity, I am for smaller game. Returning to the wonder of what is in topsoil might be a starting point. This is to say that a key response to the Baconian hubris of altering nature to our comfort is *humility*. Returning to actual practices of union between nature and human being begins not with erecting the next iconic building or landscape signaling "sustainability." It rather begins with reconfigurations of the heart at the individual and subjective levels. Below I suggest retrieval of two points of view toward this end. One is from the Chinese Neo-Confucian philosopher Zhu Xi 朱熹 (1130–1200 CE), and the other is from St. Paul (d. 64 CE). What they have to say about nature provides a theoretical framework for humility. Specifically, what they have to say about nature places human being *in proportion* to the natural cosmos as a single distribution. And this resonates with Leopold's land ethic. In this chapter, by the word "nature" I have in mind the single biotic unity envisioned by Leopold in his land pyramid. Bacon and his ideas radically segmented this pyramid into a mechanistic kit of parts. In this process, he removed human being out of the weave of its original organic and natural distribution.[9] After a brief overview of Bacon's hubris, I show how Zhu Xi's and St. Paul's views both foster practices of humility more compatible with Leopold's model.

FRANCIS BACON

Francis Bacon (1561–1626) stands at the headwaters of The Enlightenment, the single most influential intellectual development since the advent of

Christianity in the early centuries of the Common Era. A devout Anglican, Bacon propounded a vision of nature nevertheless at odds with how Christian commentators in the centuries previous to him understood the received teachings of the Christian Bible. The point is that our contemporary and largely normative tendency to blame the Christian worldview for degradations of the natural environment traces back to Bacon, not to earlier understandings of the natural order of things from Clement (first century CE) to at least Thomas Aquinas (1225–1274) if not afterward. Lynn White's influential essay "The Historical Roots of Our Ecologic Crisis," first published in 1967, is one example of this misdirected critique. The following excerpts from White's essay make his antagonism to Christianity clear:

> What did Christianity tell people about their relations with the environment? . . . Christianity made it possible to exploit nature in a mood of indifference to the feelings of natural objects . . . as we now recognize, somewhat over a century ago science and technology - hitherto quite separate activities - joined to give mankind powers which, to judge by many of the ecologic effects, are out of control. If so, Christianity bears a huge burden of guilt.[10]

White's animus is echoed by other influential commentators such as Ian McHarg[11] and Leo Marx.[12] But "exploiting nature in a mood of indifference to the feelings of natural objects" is very much thanks to Bacon, not thanks to Biblical revelation. In Bacon's day, the cultural gaze was turning toward *objective measures* to comprehend every possible experience. Just prior to Bacon, Luther's Protestant Reformation (1517) reframed Catholic mystical liturgy into specific *propositions* of faith (in some ways not unlike scientific equations) that can be proliferated in print—enabled by a new technological innovation, the printing press. And just after Bacon, Renee Descartes's (1596–1650) famous "I think therefore I am," held that we must doubt every experience of the senses until it can be comprehended by the experimental powers of human reason. Along these lines, Bacon promoted an inductive approach to observation of natural phenomena. For Bacon, this labor of inductive thinking is "extracted not only out of the depths of the mind, but out of the very bowels of nature."[13] Wording such as *exclusion, extraction, inquisition* (of things), *artificially devised* instruments for experimentation, mathematics as "a *machine* besides you" to help comprehend nature; all of these oppositional wordings abound in Bacon's writings.[14]

But Bacon himself thought he was doing the Lord's work. Of his *Description of a Natural and Experimental History*, Bacon said:

> I want this primary history to be compiled with a most religious care, as if every particular were stated upon oath, seeing that it is a book of God's works and

(so far as the majesty of the heavenly may be compared with the humbleness of earthly things) a kind of second Scripture.[15]

Thinking that he can add a "second Scripture" to the Biblical canon hints at Bacon's hubris. His attack on nature—with religious care—begins with a rejection of established ways to contemplate God. Bacon received from medieval scholasticism (particularly its Aristotelian tinge through Aquinas) what he held was a static body of speculative knowledge. Bacon held this knowledge to be useless for developing new (read: *useful; technological*) knowledge about nature. "The entire fabric of human reason which we employ in the inquisition of nature is badly put together."[16] What is needed is "a total reconstruction of sciences, arts, and all human knowledge."[17] In Bacon's mind, this requires overcoming four different "idols." Let me briefly summarize these four idols and explain how Bacon's critique of each inadvertently impacts our current views of sustainability and ecology.

The first to be dispensed with are *idols of the tribe*.[18] This is knowledge we derive from our senses. The senses are "a false mirror"; they cannot reach indubitable conclusions. A few decades later, Descartes' skepticism will echo this view: "neither our imagination nor our senses could ever assure us of anything."[19] But here is the problem. Rejection of the senses results in removal of the embodied human being *in* what Leopold calls the land pyramid: "a fountain of energy flowing through a circuit of soils, plants, animals."[20] In this flow, embodied knowledge is largely aesthetic (felt) knowledge, rather than empirically measurable knowledge. Ridding the senses in seeking exclusively measurable knowledge eliminates the aesthetical dimension inherent in holistic engagements with nature. Take away the aesthetics of the cosmos and nature becomes utilitarian; we can *do* things to nature to achieve human ends. For instance, Leopold notes that, in the name of economic utility, even the "conservation esthetic" serves a scenographic agenda.[21] National parks; hunting-grounds artificially cordoned off; maximizing views along highways to scenic locations. All of these attempts to put humans "back in nature" only increase separations from nature. These are some of the complications that arise with Bacon's rejection of the senses.

Second to go are *idols of the cave*. Here, Bacon rejects the validity of individual human experience: "the spirit of man (according as it is meted out to different individuals) is in . . . fact a thing variable and full of perturbations . . . governed as it were by chance."[22] True knowledge must be blind to the "perturbations" of individual preferences, and the personal histories shaping those preferences. In architectural theory, one example of this objectification is the New Urbanist approach to town planning, which imposes on any neighborhood plan a recipe of town greens, mixed used buildings, front porches instead of garages, and so on.[23] Elizabeth Plater-Zyberk, the designer of well-known New Urbanist communities such as Seaside, Florida, said this: "By providing a full range of housing types and workplaces . . . the bonds of an authentic community are formed."[24] But this is a startling claim. It assumes that authentic community can be achieved by figure-ground patterns of material objects arranged just-so. Not a word about the individual human beings who comprise those communities; their stories; their histories; their hopes and aspirations. Just the mixed-use buildings will do. Few readers see that this objectification is an outgrowth of Bacon's elimination of individual human experience (or "perturbations") in pursuit of useful knowledge. But it is.

Third are *idols of the market*. These amount to "the commerce and consort of men . . . for it is by discourse that men associate, and words are imposed according to the apprehension of the vulgar."[25] His elitist tone aside, Bacon is after the slipperiness of words: "words plainly force and overrule the understanding . . . and lead men away into numberless empty controversies." It is because of the slipperiness of words that Bacon rejects the knowledge of the past: "what should be said of the scholastics? Their subtleties were spent on words or . . . on common notions, not on things *or on nature*; they were quite useless in their origin and in their consequences" (italics added).[26] Here, we see the early beginnings of the idea that words are undependable signs, later taken to its logical end by Jacques Derrida. My focus here is not on the linguistic turn, which lies at the philosophical core of structuralist and post-structuralist "incredulity towards metanarratives"[27] (because the words that construct them are arbitrary). I only note that the unintended consequence of Bacon's skepticism against words is the erasure of *community*, that is, the "commerce and consort of men . . . for it is by discourse that men associate." When common sense meanings of words are in doubt, sense of community evaporates. In stark contrast, Leopold says this:

> All ethics . . . rest upon a single premise: that the individual is a member of a community of interdependent parts. His instincts prompt him to compete for his place in that community, but his ethics prompt him also to co-operate . . . The land ethic simply enlarges the boundaries of the community to include soils, waters, plants and animals, or collectively: the land (172).

This is a different view of nature. It is not a nature to harness for production; it is a nature in which humans participate in-step with the biota of the cosmos; and all of this lived as a communal enterprise. The Baconian view would have none of it.

Fourth are *idols of the theater*. By this Bacon rejects the history of ideas: "there are idols which have immigrated into men's minds from the various dogmas of philosophies, and also from wrong laws of demonstration."[28]

They result merely in "stage plays . . . after an unreal or scenic fashion." And so we must be on the lookout against "more plays of the same kind . . . yet to be composed in like artificial manner."[29] It is a common human foible: get rid of the past so we can invent a new future. As I write, all sorts of historical monuments and works of art are removed from their public settings because they do not meet present-day cultural imperatives. I suggest that this rejection of history is itself not congruent with a sustainable agenda for human community, which calls for humility, one that learns from the past so as to grow into the future. The rejection of the past to pursue the illusion of present-day "certainty"—Bacon was a trailblazer in this way of erasing the past.

For the readership of this book, it is probably not necessary to elaborate on just how influential Bacon's ideas were for the subsequent development of science, the advent of the Industrial Revolution, and the condition known as "modernity." Bacon was at the headwaters of the modernity that European ideas of the seventeenth and eighteenth centuries bequeathed to the world. It was Bacon who first separated the human being as a thinking unit unto itself, divorced from the fabric of the cosmos. And from that nowhere position, the human technological mind *dictates* how nature is to be analyzed, managed, and exploited for purposes of utilitarian comfort. The irony is as follows: At least in my domain of architectural education and practice, Baconian hutzpah is quite evident in our sustainable design "imaginary."[30] Take for instance this boastful statement from *the Leadership in Energy and Environmental Design* (LEED) website: "LEED v 4.1 is bigger, stronger, bolder . . . it is more inclusive with updated referenced standards and allows projects to earn LEED points through building performance monitoring."[31] Is this really a spirit of sustainability? What we have here is the same dominance over nature Bacon promoted, now redirected in a can-do triumphalism to "save" nature. In either manipulating nature for the sake of extraction, or in "sustaining nature" by human hutzpah, we humans are masters of the universe. There is a loss of any sense of participatory *proportionality*.

In what follows, I suggest that both Zhu Xi and St. Paul offer views of the cosmos that provide more robust theoretical bases for situating human being *in* the larger system of nature; and this *in proportional measure*. Because their systems do not segregate human being from nature, retrieval of these two points of view supports Leopold's land ethic. Again, recalling these two earlier views is not a proposal for change on a global scale. It simply calls for humility at individual—which is to say, at small and at personal—scales. After all, philosophy both East and West prior to the Enlightenment had more of a moral agenda: Pursuit of truth was linked to cultivation of virtues at the individual level. As we will see, this moral element shines through in both Zhu Xi and St. Paul. Indeed, prior to Bacon, alterations of nature by

large-scale technological impositions would have seemed odd at best, perhaps even *im*moral.

HUMILITY BY PROPORTION (1): ZHU XI
AND THE INVESTIGATION OF THINGS

The most significant impact the influx of European ideas had on China was
the bifurcation of human being from the natural order of things. Chinese
philosophy never envisioned a chasm between human being and "everything
else" akin to Descartes' divide between *res cogitans* (the thinking thing) and
res extensa (material things with no capacity to think). Bacon's "inquisition
of nature" is the former interrogating the latter. While traffic existed between
China and the West for centuries prior, it was the Opium Wars of the nineteenth century (specifically, in 1839–1842 and again in 1856–1860) that
thrust the power of industrial mechanization upon a culture grounded for millennia in an agrarian way of life. Gunboats, armaments, railroads, and other
mechanized equipment rudely awakened Chinese civilization to how "nature"
can be reconfigured into utilitarian commodities that can wreck tremendous
damage—and also fabricate objects stirring tremendous desire. I have written
elsewhere on how industrialization in China created the curious juxtaposi-
tion between traditional and modern worldviews embedded in the "Chinese
modernity" that impacts China's architecture and urban environments even
to this day.[32] Not all of it is good. Rural migration to today's economically
booming cities in China has resulted in the loss of nearly 920,000 traditional
villages nationwide in the last fifteen years, and the decline continues.[33]
    Zhu Xi lived in the Song dynasty (960–1279 CE), a time of enormous
advancement in Chinese civilization. In the realm of ideas, it was also a time
of retrieval. By the Song period, the impact of Buddhism, which entered
China along trade routes to the west during the Han Dynasty (206 BCE–220
CE), had revolutionized Chinese culture. By Zhu Xi's day, Confucian literati, even while pushing back against Buddhist doctrines and practices (e.g.,
young people leaving their parents to join Buddhist sanghas; a practice abhor-
rent to Confucianism), had themselves been imbibing from Buddhism simply
by living in the culture. So "retrieval" of past ideas for Neo-Confucian thinkers leading up to Zhu Xi necessarily included Buddhist (and Daoist) ideas
intermixed into their formulations. Largely because of Buddhist influences,
Zhu Xi's Neo-Confucianism contains a metaphysical emphasis not found in
the early Confucianism of the *Analects* or even the *Mencius*.
    So much for background. In what follows, I will first consider several
Chinese philosophical terms often translated "nature"; all of these terms
informed Zhu Xi's philosophy. These terms underline the nonseparation

between human being and cosmos in general—*even while this seamlessness organized around a sense of proportion in which (small) human being participates in (large) cosmos.* I will then give one key example from Zhu Xi that profoundly illustrates this seamless continuity in his thought. I then suggest how Zhu Xi's retrieval of an ancient term, "the investigation of things" 格物 (*ge wu*), illustrates the Neo-Confucian cultivation of humility vis-à-vis the natural world.

We can begin with nature understood as *xing*, 性, referring to the inherent property of a thing. This comes closest to the Aristotelian view that each thing has its own *telos*. But there is no necessary directionality in this Chinese term so that all things arrive at an ideal polity, as Aristotle would have it. Indeed, the Yangist school in the ancient China eschewed communal engagement so that *xing* would not be clouded by (what Rousseau might call) the conventions of social contract.[34] Some commentators have suggested that the ancient Chinese view of all reality was that of a single fabric cut into pieces (by words) to denote meanings.[35] These pieces (*fen*, 分) are the *portions* (read: proportion) that comprise people and "the myriad things." Donald Munro has noted that, in this mindset, to be morally "wrong" is to trespass beyond one's *fen*.[36] Such an understanding of all of reality as a single distributed field was sufficient to either build philosophies of social engagement (Confucianism) or philosophies of social nonengagement (Daoism). The point for us is that both views saw human being in proportional fit to the cosmos; it is just a question of how, in one's smallness, one participates in the larger whole, so as to cultivate one's *xing*. No Baconian hubris here.

Another early term for nature is *zi ran*, 自然. Importantly, this is the same term for *spontaneity*. Western minds should pause to grasp the import of this equivalence in formulating prescriptive approaches to conservation: As soon as there is deliberation of any kind, we depart from the spontaneity of *zi ran* = nature. This is the basis for early Daoism's avoidance of social conventions. The *Daodejing*, dating from the Warring States Period (476–221 BCE) if not before, says this: "Thus (the sage) supports all things in their natural state but does not take any action."[37] This is because the sage's nature (*xing*) is completely at-one with "the natural state" (*zi ran*). The point is that the sage does not take any *deliberative* action, the exercise of which separates from natural spontaneity. This is why, I often render the term for not taking any deliberative action (*wu wei*, 无为) as "letting nature flow through you."[38]

And letting nature flow through you is a proportional operation. When Zhu Xi says the following (he is citing from an earlier Neo-Confucian thinker), he was clearly incorporating this Daoist view: "The constant principle of the sage is that his feelings are in accord with all creation, and yet he has no feelings of his own."[39] This is a statement of proportion: The sage is effective precisely thanks to his own ontology being so small in relation to the operations of nature. Moral propriety comes by deferring to those larger operations. This kind of sagehood is impossible without humility.

Third, the Confucian School prized *tian*, 天. Often translated Heaven, this is not the Christian idea of a heavenly order distinct from "the kingdom of this world." *Tian* is also translated sky, but it is more than just the blue above. *Tian* is something like the complete natural order of things, but with moral volition. Indeed, when the Zhou polity overthrew the early Shang in 1046 BCE, their rationale for victory was the "the mandate of heaven (*tian*)." So there is a way that the total order of things—nature—not only includes human beings in its organic workings; this natural order of things exercises a volitional moral force. What is more, this moral force can be understood as a humaneness (*ren*, 仁) that permeates the natural order (*tian*). This is why Zhu Xi can say, "if we can overcome and eliminate selfishness and return to the Principle of Nature (*tian-li*), then the substance of this mind (that is, the mind of humaneness) will be present everywhere."[40] Smallness of scale on the part of the human being in relation to *tian*, and the humility this position entails, is also evident here.

Fourth, the most important term Neo-Confucianists adapted from the past is principle (*li*, 理). A relatively minor term in ancient ideas, Zhu Xi and his predecessors elevated *li*-principle to primary status by holding that all things have *li*-principle. It was a dexterous move, because another term pronounced the same way (*li*) but written differently (礼) was a key term in traditional Confucianism. This other *li* term denotes all of the social rituals one participates in to fulfill the demands of one's social *fen* (portion)—and hence to fit proportionally into the fabric of society. So here is the play on words: For one to be true to *li*-principle, one must cultivate *li*-ritual. In Zhu Xi, this is a lifelong practice with metaphysical scope. Indeed, *li*-principle is simply the Great Ultimate itself, which is "merely the principle of heaven and earth and the myriad things."[41] Munro cites Zhu Xi:

> In looking, hearing, speaking, and acting, when one does nothing to (*li-*) principle, this is acting according to the rules of (*li-*) propriety [that is, *li*-ritual]. The rules of propriety are principle. Whatever is not in accord with heavenly principle is selfish desire.[42]

For the thoughtful individual in daily life, then, cultivating *li*-ritual leads to enlightenment in *li*-principle. By doing so, the individual fits in proportionately to the *li*-principle of the Great Ultimate.

Here, then is the example from Zhu Xi alluded to earlier. Consider the moon and reflections of moonlight on bodies of water. For Zhu Xi, the *li*-principle of the moon is the same as the *li*-principle of those reflections. It is not that the moon is "the actual thing," while the reflections are some sort of second-order representation of the moon. The moon and reflections both participate in the same *li*-principle of the moon, which in turn participates in the *li*-principle of the Great Ultimate. "It cannot be said that the moon has been split."[43] This comes from a mind, and an overall ideological tradition, that entertained no Baconian divide between human reason and the natural world.

To further propound this seamless distribution, Zhu Xi looked to the ancient document *The Great Learning* (Warring States 476–221 BCE; perhaps earlier). That document offers a window into seeing how the human family and the cosmos were always viewed as a seamless whole. *The Great Learning* holds that if one knows how to regulate the family, "there will be peace throughout the world."[44] Munro:

> As a persuasive device, the image of the family writ large in nature is intended to have an emotional impact on those who accept it. They should react with family-derived sentiments (love, respect, obedience) to the heavens and the earth, the animals, and the trees in nature."[45]

Munro's point certainly underlines how *The Great Learning* envisions human beings living in proportion to the cosmos, in which "the myriad things"— the animals, the trees in nature, and so on—fit together proportionately on familial terms. Space does not allow for elaboration of the step-by-step prescriptions for cultivating the family, leading to peace in the world. But we can list them: To regulate a state, first regulate your family; cultivate your personal life; rectify your mind; make your will sincere; extend your knowledge; investigate things (*ge wu,* 格物, see below). When these are done: knowledge is extended; the will becomes sincere; the mind is rectified; personal life is cultivated; the family is regulated; all the world will be at peace. The fulcrum of this process—the key—is the *investigation of things*. The Neo-Confucianist project, then, was to recognize the essential compatibility between human life and the natural order (by means of the common denominator of *li*-principle in all things), and to cultivate this union by the investigation of things.

What we see here is a practical way of living-in-the-world that is transferable to praxis in the land ethic. How we look, how we hear, how we speak and how we act; the cultivation of a land-ethical way of life begins on the individual level, in humble submission to the orderliness of "the myriad things." At minimum, it rejects the tendency of the "professional" enactments of initiatives for sustainable design, which is that the prescriptions are always *for others*. True sustainability begins here, locally, at home. It begins with me as I look for ways to humbly participate as-one with the organic workings of the myriad things.

PROPORTION AND HUMILITY (2):
ST. PAUL AND RECYCLING

One way to be as-one with the organic workings of the myriad things is to participate in recycling. Now, with St. Paul as the focus of this section, of course I have a particular notion of recycling in view. I suggest that all acts of recycling—of plastics, glass, cardboard, and so on—are echoes of *death and resurrection* inherent in life processes permeating the natural order. Implicit in the citation from Berry about topsoil is this: death and resurrection is the *complete* rationale for recycling; this because death and resurrected life is the *organic* rationale for recycling. Conventional understandings of recycling typically take care to transfer one form of embedded energy to another form of energy in a balance-sheet view of the cosmos. It follows the standard rubric of "meeting the needs of the present without compromising the ability of future generations to meet their own needs."[46] But leaving no less for subsequent generations doesn't seem to match the *nature* of the nature we wish to sustain. The nature we wish to sustain *explodes* with life. Rather than a calculated net-zero accounting of things, the death and resurrection in topsoil observed by Berry leaves the next iterative cycle with *abundance*. Is this not the joy of participation in the natural world? Nature rejuvenates, it heals, it reproduces beyond expectation. This fecundity goes to the core of Leopold's ecological rationale for the land pyramid (italics mine):

> Land, then, is not merely soil; it is a fountain of energy flowing through a circuit of soils, plants, and animals. Food chains are the living channels which conduct energy upward; death and decay return it to the soil. The circuit is not closed; some energy is dissipated in decay, some is added by absorption from the air, some is stored in soils, peats, and long-lived forests; but it is a sustained circuit, like a slowly *augmented revolving fund of life.*[47]

On Pauline terms, the essence of this "revolving fund of life" is *incarnated divine life.* In today's post-Christian culture, St. Paul's claims are rarely mined for what they offer for various domains of academic discourse. Of course theologians talk to each other about the Bible and publish books that usually amount to "preaching to the choir." But if we get beyond the strictures of "religion as a private matter," the Christian Scriptures offer a wealth of theoretical material for informing current challenges in environmental degradation. Ellen Davis has persuasively shown how the Old Testament books would be almost incomprehensible without an awareness of the agrarian logic inherent in their content;[48] hence, those sources are invaluable for working alongside efforts to instill Leopold's land ethic. And in the New Testament, St. Paul writes of an incarnation of God as human being that renders the entire cosmos *humane*: "For from him and through him and to him are all things."[49] "He who descended is the one who also ascended far above all the heavens, that he might fill all things."[50]

Just like Zhu Xi, St. Paul's vision is of a single fabric of the cosmos. There is no divide between empirical versus immaterial, qualitative verses quantitative, personal versus public, urban versus rural, and other bifurcations of current social construction. In contrast to these divisions, the organizing proportionality of incarnation is *humanness*. Just like Zhu Xi, St. Paul discerns a humanity that weaves through the fabric of the cosmos. For Zhu Xi, as noted above, it is *ren*, 仁. The pictograph has the "human" radical (亻) embedded in it, and is often translated "humanity," or "benevolence." In the Chinese case, the humanity-benevolence in the cosmos is something like an inherent virtuous presence which, when cultivated by *li*-ritual to touch on the *li*-principle in all things, ushers forth in *proportionally distributed* humane social relations—all of it understood in familial terms. In fact, the last prominent Neo-Confucian thinker, Wang Yangming (1472–1529 CE), held that

> The great man regards the world as one family and the country as one person. . . . That the great man can regard Heaven and Earth, and the myriad things as one body is not because he deliberately wants to do so, but because it is natural to the humane nature of his mind.[51]

In St. Paul's case, this humaneness is God incarnated as a human person, the Christ:

> He is the image of the invisible God, the firstborn of all creation. For by him all things were created, in heaven and on earth, visible and invisible, whether thrones or dominions or rulers or authorities—all things were created through him and for him. And he is before all things, and in him all things hold together.[52]

To most academic minds, this sounds implausible. But why must it be? It is not my domain, but there is substantial literature in the sciences propounding *the anthropic principle*, to wit, that we humans can comprehend the cosmos precisely because there is something about its composition conducive to *human* comprehension. There is a like-unto-like quality about our place in— and communion with—the cosmos. And just like Zhu Xi, the human being is small in relation to the immense workings of the organic and humane cosmos. But the key is that he or she *participates* in the organic percolations of the natural order of things. And that organic percolation is analogous to the death and resurrection of the Christ.

What is more, St. Paul deploys agricultural analogies to usher in his point about participation by death and resurrection. He was not the innovator in these analogies. The Christ himself said this: "unless a grain of wheat falls into the earth and dies, it remains alone; but if it dies, it bears much fruit."[53] This is the operative principle of sustainable abundance that rhymes essentially with Leopold's land ethic, and with Berry's cautious observation that the life and death he sees in topsoil just might be miraculous. St. Paul used cosmic-agricultural terms to describe this kind of recycling:

> There is one glory of the sun, and another glory of the moon, and another glory of the stars; for star differs from star in glory. So is it with the resurrection of the dead. What is sown is perishable; what is raised is imperishable. It is sown in dishonor; it is raised in glory. It is sown in weakness; it is raised in power.[54]

What I am suggesting is that there is enormous analogical power—the Latin *analogia* is translated *proportion*—in seeing the entirety of the natural order of things as an analogous weave of being, in which human being, and all other beings, participate (analogously) with the Being that enables the entirety of the cosmos. On the Scriptural accounting, that Being is God; in its humane mode, it is the incarnated Christ now distributed and filling all things (or, on Zhu Xi's terms, filling the myriad things).

Analogy in this sense needs further elaboration. For one and a half millennia prior to the events of the Enlightenment, human being was not seen as dichotomous from the natural order. This means the only way to comprehend life-conditions was to see analogy not as a comparative device between two dissimilar things (e.g., he has feline features like his cat), but rather as a statement of actual conditions. When the Christ said "unless a grain of wheat fall into the ground and dies, it remains alone; but if it dies, it brings forth much fruit,"[55] he was not *likening* himself to a grain of wheat. In an actual-analogical sense, he *is* the grain of wheat that falls into the ground and dies. And insofar as what he said takes on the proportions of an operational principle for the cosmos, we also are not "like" grains of wheat; we also *are* grains of wheat—analogically comprehended. If we fall into the ground and die (analogically), we are in-step with a cosmos that recognizes this humility. The outcome is an explosion of life when night turns into day, when winter turns into spring; when putting others first ahead of myself in the land pyramid turns into healthy community.[56]

CONCLUSION

Counter to Baconian hutzpah, what Zhu Xi and St. Paul both tell us is that, to be sustainable, we must see our place in the cosmos not as beings cut out of the natural fabric of things and lording it over them, but rather as integral to the cosmos. For both thinkers, this requires humility.

If I had more space, I would further argue that Zhu Xi's and St. Paul's views can be appreciated through the lenses of *beauty* and *sublimity*, respectively. Let me touch on these briefly by way of conclusion. Of course beauty and sublimity are aesthetic categories that themselves came out of the European Enlightenment. Put another way, as Bacon's technological "extractions" of nature began to bear fruit in providing more leisure time for the aristocratic classes, the idea of *fine art* emerged—that is, objects of appeal which precisely had *no* utility in a Baconian world bent on utilitarian subjugations of nature. This inconvenience aside, beauty and sublimity, as categories of aesthetic human experience, remain useful for further analyses of Zhu Xu and St. Paul.

Experiences of beauty entail *union* (variously theorized) between one's subjective sensibilities and external objects, whether of nature or of art. An early theorist in these matters, Francis Hutcheson (1694–1745), for instance, held that humans had an innate "sense of beauty," which immediately connects with objects that are beautiful.[57] For his part, Immanuel Kant (1724–1804) was so taken by the apparent purposiveness behind nature's orderliness that he described aesthetic enjoyment of the natural world has nature embracing the human being as a "member."[58] In Zhu Xi, we have a trace of beauty in this sense. Further research can develop on Zhi Xu's ideas in context of these formulations for the beautiful, to wit, that *li*-principle within me rhymes with *li*-principle in the myriad things.

While experiences of beauty derive from similarity between human being and encounters with external objects (whether in nature or art), experiences of sublimity emphasize dissimilarity. The cosmos presents me with a natural array so immense, so overwhelming, that my pleasure derives from amazement that I can even survive such a powerful display. In his accounting of the mathematical sublime, Kant holds that such immensity stirs ideas of the *absolutely large*.[59] And this experience is pleasurable. St. Paul's writings are filled with this awe of the absolutely large, because his vision is of a cosmos in which God comes as an incarnated human being, to fill the very immensity of it. "For in Christ all the fullness of the Deity lives in bodily form, and in Christ you have been brought to fullness. He is the head over every power and authority."[60]

In both Zhu Xi (the beautiful) and St. Paul (the sublime), the natural order includes the human being as a small participant in its enormity. What Bacon did was to erase this sense of proportion. If anything, in Bacon, human hubris becomes large and the cosmos becomes small because it is seen as mere matter at our disposal. Such a distortion of scales greatly reduces experiences of the beautiful and the sublime. Such a distortion of scale erases the possibility of the kind of humility envisioned by both Zhu Xi and St. Paul. And humility is an important ingredient in returning to a sensible—and sensibly ecological—membership in the land pyramid of which we are a part.

NOTES

1. In referring to the human ontology in general, I use the singular (e.g., human being) in this chapter.

2. Aldo Leopold, "The Land Ethic," in *A Sand County Almanac & Other Writings*, edited by Curt Meine (The Library of America, 2013), 171–189.

3. Ibid., 180–181. The definition of land pyramid is cited later in Leopold's chapter.

4. Charles Taylor, *A Secular Age* (Harvard University Press, 2007).

5. Ibid., 539–594.

6. For a summary of Taylor's extensive work, see James K. A. Smith, *How (Not) to be Secular* (Grand Rapids: Eerdmans, 2014).

7. Taylor, op. cit., 549.

8. Wendell Berry, "Two Economies," in *Every Man an Artist: Readings in the Traditional Philosophy of Art*, edited by Brian Keeble (World Wisdom, Inc., 2005), 192.

9. Allen Verhey actually discerns no less than 16 different meanings for the word "nature." It is my view that this confusion itself comes out of Bacon's segmentation of human being from the natural world. See *Nature and Changing It* (Grand Rapids, MI: W.B. Eerdmans, 2010), 1–12.

10. Lynn White, "The Historical Roots of Our Ecologic Crisis," *Science*, Vol. 155, No. 3767 (March 10, 1967), 1203–1207.

11. "The creation story in Judaism was absorbed unchanged into Christianity. It emphasized the exclusive divinity of man, his God-given dominion over all things and licensed him to subdue the earth . . . . From early, the faintly ridiculous beginnings when a few inconsequential men proclaimed their absolute supremacy to an unhearing and uncaring world, this theme has grown." Ian McHarg, *Design with Nature* (New York: Doubleday, 1971), 26–27.

12. "The utilitarian bias was buttressed by the strong Protestant sense of the natural world as lawless, unredeemed, satanic" Leo Marx, "The American Ideology of Space," in *Denatured Visions: Landscape and Culture in the Twentieth Century* (Museum of Modern Art, 2003), 64.

13. Francis Bacon, "The Great Instauration," in *The New Organon and Related Writings*, edited by Fulton H. Anderson (Indianapolis and New York: Bobbs-Merrill Company, 1960), 21.

14. Ibid., see 18, 22, 27.

15. Francis Bacon, "Description of a Natural and Experimental History" in Ibid., 282.

16. Bacon, "The Great Instauration," in Ibid., 3.

17. Ibid., 4.

18. Bacon, "The New Organon" section XLI, in Ibid., 48.

19. Renee Descartes, Discourse 4, in *Discourse on Method and the Meditations*, translated by F.E. Sutcliffe (Penguin, 1968), 58.

20. Leopold, op. cit., 181.

21. Leopold, "The Conservation Esthetic," in op. cit., 143–151.

22. Bacon, "The New Organon," section XLII, in Ibid., 48–49.

23. "New Urbanism: Creating Livable Sustainable Communities. http://www .newurbanism.org/sustainability.html. Accessed March 19, 2019.

24. Elizabeth Plater-Zyberk, quoted in Daralice Boles, "Reordering the Suburbs," *Progressive Architecture* 5 (1989): 78–91.

25. Bacon, op. cit., XLIII, 49.

26. Ibid., CXXI, 110–111.

27. Of course this is Jean-Francois Lyotard's term in *The Postmodern Condition: A Report on Knowledge*, translated by Geoff Bennington and Brian Massumi (Manchester University Press, 1984).

28. Bacon, "The New Organon" section XLIV, in Ibid., 49.

29. See for instance Roger Kimball, "Shall We Defend our Common History? In *Imprimis*, Volume 48, Number 2, February, 2019. https://imprimis.hillsdale.edu/shal l-defend-common-history/. Accessed March 19, 2019.

30. "Imaginary" is a term from Charles Taylor. It is somewhat similar to "worldview." See James K. A. Smith's explanation of this term in *How (Not) to be Secular* (Grand Rapids: Eerdmans, 2014), 24–26.

31. "LEED v4.1, Who knew '.1' could deliver so much." https://new.usgbc.org/ leed-v41. Accessed March 6, 2019.

32. David Wang, *A Philosophy of Chinese Architecture Past, Present, Future* (Routledge, 2017), 85–109.

33. Professor Liu Xiaohu, WeChat correspondence with author, March 28, 2019. He shares this article: "Chinese ancient villages are in urgent need of rescue and protection," 中国古村落亟待抢救保护, Zhōngguó gǔ cūnluò jídài qiǎngjiù bǎohù, in *China City News*, December 18, 2017. http://paper.people.com.cn/zgcsb/html/20 17-12/18/content_1824614.htm. Accessed March 28, 2019.

34. A.C. Graham, *Disputers of the Dao* (Chicago and La Salle, IL: Open Court, 1989), 56–59.

35. Chad Hanson, *Language and Logic in Ancient China* (Ann Arbor: University of Michigan Press, 1983), 30–54.

36. Donald Munro, *Images of Human Nature: A Song Portrait* (Princeton University Press, 1988), 50.

37. *Daodejing* 64, in Wing-Tsit Chan, "The Natural Way of Lao Tzu," in Wing-Tsit Chan, *A Sourcebook in Chinese Philosophy* (Princeton University Press, 1963), 170.

38. David Wang, "Forward," in *The Tao of Architecture*, by Amos Ih Tao Chang (Princeton University Press, 2017), viii.

39. Zhu Xi (Chu Hsi), in Chan, *A Sourcebook*, op. cit., 643. The citation is from Cheng Hao (1032–1085).

40. Ibid., 594.

41. Ibid., 638.

42. Cited in Munro, Ibid., 55. This reference is identified by Munro as "*I-shu*, 15.1b."

43. Zhu Xi: "Fundamentally there is only one Great Ultimate, yet each of the myriad things has been endowed with it and each in itself possesses the Great Ultimate in its entirety. This is similar to the fact that there is only one moon in the sky but when its light is scattered upon rivers and lakes, it can be seen everywhere. It cannot be said that the moon has been split." Ibid., 638.

44. "The Great Learning," in Ibid., 94–95.

45. Munro, op. cit., 54.

46. William McDonough & Partners, *The Hannover Principles: Design for Sustainability,* 1992. https://www.mcdonough.com/wp-content/uploads/2013/03/Hannover-Principles-1992.pdf. Accessed March 22, 2019. The citation is actually from the 1987 United Nations Brundtland Commission's definition for sustainable design.

47. Leopold, "The Land Ethic," op. cit., 180–181.

48. Ellen Davis, *Scripture, Culture & Agriculture* (Cambridge University Press, 2010).

49. Romans 11.36.

50. Ephesians 4.10.

51. Wang Yangming, "Inquiry on the Great Learning," in Chan, op. cit., 659.

52. Colossians 1.15–17.

53. John 12.24.

54. 1 Corinthians 15.41–43.

55. John 12.24.

56. Analogy in the sense I am using here has a technical name in theology: *analogia entis,* or analogy of being. This topic is too broad to explain here. My *Architecture and Sacrament: A Critical Theory* (Routledge, 2020), addresses *analogia entis* throughout its seven chapters.

57. Francis Hutcheson, *An Inquiry into the Original of our Ideas of Beauty and Virtue* (1729), Section II.5. https://oll.libertyfund.org/titles/hutcheson-an-inquiry-into-the-original-of-our-ideas-of-beauty-and-virtue-1726-2004. Accessed April 20, 2019. "In every Part of the World which we call Beautiful, there is a Uniformity amidst almost infinite Variety. Many Parts of the Universe seem not at all design'd for the use of Man; nay, it is but a very small Spot with which we have any acquaintance . . . and yet as far as we can by Sense discover, or by Reasoning enlarge our Knowledge, and extend our Imagination, we generally find their Structure, Order, and Motion, agreeable to our Sense of Beauty."

58. Immanuel Kant, *Critique of Judgment*, Sec 67, Translated by Warner S. Pluhar. (Indianapolis: Hackett Publishing Co., 1987), 380; pagination is the marginal pagination carried over from the original edition.

59. Ibid., Sec 25. There is much more here that I cannot elaborate on due to space. For example, Kant's theory holds that it is not the external immense object that is the primary reason for my experience of sublime pleasure, but rather that my internal faculties are overwhelmed by a sense of absolute immensity.

60. Colossians 2.9–10.

Chapter 2

Old Dreams Retold*
Lu Xun as Mytho-Ecological Writer
Ban Wang

Regarded a champion of the May Fourth ideals of democracy and science and a relentless critic of Chinese tradition, Lu Xun is less known as a prescient critic of Enlightenment, science, and rationality. Just as Walter Benjamin and Theodor Adorno attempted to invoke utopian images of the reconciliation of humans and nature from premodern and precapitalist pasts, Lu Xun conjured up images of the ancient world, where rural folks lived in reciprocity with nature, worshiped supernatural beings, and observed time-honored rituals. Lu Xun's turn toward the past is paradoxical and provocative, and is evident in his defense of "superstition." In "Po esheng lun" 破惡聲論 (Toward a refutation of malevolent voices), an essay written in 1908 while he was a student in Japan, Lu Xun linked the myth of progress and modernity to a chorus of "malevolent voices" embodied by a bunch of technocratic elites and modernizers he called *weishi* 偽士 (hypocrite gentry). In the name of enlightenment, progress, and rationality, the hypocrite gentry dominated public discourse, fetishized science and technology, and sought power, prestige, and profit by trashing Chinese "superstition." Lu Xun hurled critique at this elite by claiming that it was urgent to "rid of ourselves of this hypocrite gentry; "superstition" may remain" (*weishi dangqu, mixin ke chun* 偽士當去，迷信可存).[1]

Using the past to critique modernity is indebted to the Romantic tradition. It is no wonder to see in Frankfurt School critics a similar gesture of invoking the mythical past in their critique of an Enlightenment gone awry. Water Benjamin quoted Nikolai Leskov, a Russian religious writer, to conjure up the resonant human–nature relationship embedded in the premodern life-world. In ancient times, "the stones in the womb of the earth and the planets

* Print by permission of Duke University Press of Ban Wang, "Old Dreams Retold: Lu Xun as Mytho-Ecological Writer," in *Prism*, volume 17, no. 2, pp. 225–243. Copyright, 2021, Lingnan University.

at celestial heights were still concerned with the fate of men." But modernity, with its technological exploitation, formal logic, rationality, and conquest of nature, has reduced stones to raw material and severed their millennial ties to humanity. In the new age of modernization and industrialization, stones and everything else "in the heavens and beneath the earth" have "grown indifferent to the fate of the sons of men and no voice speaks to them from anywhere." "All measured and weighed and examined for their specific weight and their density," stones are viewed as fodder for technological extraction and commodity value. They "no longer proclaim anything to us, nor do they bring us any benefit. Their time for speaking with men is past."[2]

This observation epitomizes the modern process of disenchantment. The magic conversation between humans and nature in a meaningful cosmology has been reduced to rational cognitive mappings and technological domination. The environment has become thoroughly mapped, digitized, and formalized in terms of scientific laws and as resources for industrial-technological exploitation.

This lament may contradict Benjamin's positive stance toward science and technology as a potential for avant-garde movements. In his essay The Work of Art in the Age of Mechanical Reproduction, Benjamin sought to bring media technologies such as film, photography, and print to shatter and demystify the aura of cultural authority and tradition. Film technology enlightens us by bursting "the prison-world" of our everyday perception and ideology "by the dynamite of a tenth of a second."[3] Technological reproduction could be harnessed to democratize our access to artworks and to "to pry an object from its shell, to destroy its aura"—the mythical halo surrounding traditional authority, ritual, and hierarchy.[4] In the age of imperialist wars and rising fascism, however, the dialectic wheel of enlightenment and myth turned all too quickly: No sooner did science usurp the dogma of the ancient regime and take center stage than it arrogated to itself the power of a new myth, a new God, and a new Church. The scientists and tech people ascended to a new orthodox and became a new priesthood. No sooner did technology advance human creativity than it created a new power elite bent on domination of nature and exploitation of other humans. For Benjamin, technology should ideally be used as "the natural utilization of productive forces" for improving human life. Yet the promise of progress was being "impeded by the property system" and the belligerent imperialist states. "Instead of draining rivers, society directs a human stream into a bed of trenches; instead of dropping seeds from airplanes, it drops incendiary bombs over cities; and through gas warfare the aura is abolished in a new way."[5] The techno-scientific utopia quickly turned into ruins of dystopia.

To counter this new myth, Benjamin resorted to old dreams, ancient cosmology and ecological traditions. His use of myth to critique modernity exemplifies the dialectic of enlightenment and myth expounded by Max Horkheimer and Theodor Adorno in their classic *Dialectic of Enlightenment.*

Enlightenment, armed with science and technology, made an illustrious career by dismantling tradition and myth and by unleashing intellectual freedom and scientific revolution. But the Enlightenment project quickly reversed to the roles of its enemies. Enlightenment "aimed at liberating humans from fear and installing them as masters. Yet the wholly enlightened earth is radiant with triumphant disaster."[6] It disenchanted the world, abolished myths, and became a new myth and new power in its own right.

Lu Xun also treaded on the similar dialectic ground. As an advocate of the May Fourth agenda of science and democracy, Lu Xun has been rightly regarded as a leader in the New Culture endeavor in the campaign to challenge Confucian orthodox, dispel superstition, and advance science and critical thinking. Yet like Walter Benjamin, the young Lu Xun was already skeptical of initial signs of technical rationality and material progress, and became aware early of how techno-scientific rationality and industrial modernity empowered and puffed up a bureaucratic, technocratic elite. This new elite and their ideology threatened to erase valuable strata of the traditional life-world and disenchant the earth. The suspicion of progress as regression led Benjamin and Lu Xun to turn backward to the past to sift through the ruins of folklore, myth, custom, and life-worlds to find redemptive hopes. For the intellectual context of this mythical turn in the West, Benjamin points to the period from the end of the nineteenth century to the 1930s in his essay "Some Motifs in Baudelaire" (1939). The "blinding age" of industry, technology, and consumer culture forced European philosophers to turn to "the age of myth."[7] Ancient myths contain utopian moments redeemable as "experience," a mode of feeling and relating redolent with aura and magic. Fulfillment of experience recovers traces of the reconciled life between human and nature, subject and object, and individual and community. The mythical turn sought to recover the "true" experience as opposed to the flattened structure of feeling and spiritual hollowness in the fast industrializing and consumer society. Under the rubric of "philosophy of life," the philosophers invoked "poetry, preferably nature, and most recently, the age of myth." Bergson's theory of cultural memory delves into evolutionary biology and evinces a "convergence in memory of accumulated and frequently unconscious data"—the data of premodern and prehistorical nature.[8]

The concept of myth needs a brief clarification. The idea of myth connected with primordial, premodern pasts names an ecological understanding of human–nature resonance and a holistic cosmology. It is a dream, a belief, a cultural ideal, and a vision of reality, though unverifiable by positivist means. On the other hand, the myth of enlightenment and science refers to technological rationality rooted in empiricism, and to the human hubris in the domination of outer nature and administration of inner human nature.

Confronting the triumphant discourse of progress, science, and industrialization, Lu Xun shared Benjamin's backward gaze at premodern life-worlds.

This throwback at the past challenges the blind embrace of science and modernity—the tenor of the May Fourth culture, questioning social ills of technological fetishism. The past contains a utopian future, and ancient cosmology and folk religions provide a reservoir of dream images. Like Benjamin, Lu Xun dreamed of a world where humans have magic resonance with the natural world, maintain deep communion with rocks, and are embedded in ecosystems. This wish-image is well captured by a passage in "Po esheng lun":

> In the case of China, a universal reverence for natural phenomena has always been regarded as the basis of culture. Worship of heaven and earth was the foundation of the systematic development of rituals and ceremonies. The reverence for heaven and earth, extended to the countless host of material things, provided a basis for all wisdom and moral principles, as well as our state and clan systems. In fact, the extent of its influence is immeasurable. Because of this, Chinese people have never slighted their native place, nor did social classes ever form: it was held that plants and rocks all had mysterious supernatural properties, and possessing this metaphysical significance they were viewed as different from ordinary objects. (Lu Xun 8: 27–28; Kowallis 49–50)

> 顧吾中國，則夙以普崇萬物為文化本根，敬天禮地，實與法式，發育張大，整然不紊。覆載為之首，而次及於萬匯，凡一切睿知義理與邦國家族之制，無不據是為始基焉。效果所著，大莫可名，以是而不輕舊鄉，以是而不生階級；他若雖一卉木竹石，視之均函有神軛性靈，玄義在中。

In pre-industrial rural communities, Chinese held plants and rocks in high esteem, because these objects are imbued with "supernatural properties and metaphysical significance." Emotional bonds with rocks, plants, and land are integral to attachment to the homeland through generations of peasants. Their religious reverence for natural phenomenon extends to all things, imbuing the myriad objects with aura and spirit.

By invoking the mythical past to critique the new myth, Benjamin, as Jürgen Habermas has noted, is concerned with "redeeming those utopian moments of tradition which are increasingly endangered by the oblivion of forgetting."[9] Deploying Benjamin's insights of aura, experience, and human–nature *correspondance*,[10] this chapter explores Lu Xun's recovery of the mythical and ecological elements from past life-worlds in his critique of modernity. Confronted with the triumphant discourse of progress and science in China's early project of modernization, Lu Xun sought to uncover and redeem primordial images from archaic traditions.

THE HYPOCRITE GENTRY

With the expansion of the modern system into East Asia, knowledge of modern science and technology, under the rubric of "Western learning," was introduced to China in the late nineteenth century. The educated elite embraced science and technology as the key to the power of the major Western nations and sought to emulate the techno-industrial model in order to strengthen China. Many hailed science and technology as a salvation and saw industrialization as a solution to turn the crumbling empire into a viable nation-state. This self-strengthening movement went with the pursuit of Western-style modernization, culminating in the late Qing campaign of Westernization (*yangwu yundong* 洋務運動,1861–1895). In his early writings, Lu Xun also recognized potentials of science and technology for strengthening China and for enlightening Chinese from ignorance, superstition, and benightedness. Writing about his early education, for instance, he depicted Chinese medicine as fanciful, irrational, and unscientific. One prescription for his sick father contained mysterious ingredients, such as twin crickets, aloe roots dug up in winter, and sugar cane three years exposed to frost. To Lu Xun, this "charlatan" concoction seemed to be the direct cause of his father's death. With the resolve to leave tradition behind, Lu Xun attended a modern school and received an education that exposed him to the modern disciplines of science, mathematics, geography, physiology, and chemistry.[11] His continued interest during his overseas study in Japan prompted him to write an essay "Kexue shi jiaopian" 科學史教篇 (Lessons from a History of Science). In it, Lu Xun welcomes the power of science and technology. With technological advances, mountains and rivers no long stand as barriers to transportation and communication; cures for diseases are found and plagues have been eliminated, and education and enlightenment spread. But he cautions that technological progress and material improvement should be subjected to the decree of humanity, who should harness modern technology, riding technological advances like galloping horses.[12]

In "Wenhua pianzhi lun" 文化偏置論 (On the Imbalances of Culture), Lu Xun considers the pros and cons of technological progress. Surveying the development of science and technology in the West from the Enlightenment to the nineteenth century, he notes that increased knowledge and efficient tools not only improved material life but also enlightened minds and made society more open and liberal. "When religious authority was overthrown and freedom granted to human minds, all branches of learning and technology surged with a new life." The scientific mode of thinking encouraged freedom of thought and inquiry, and discredited religious dogma.[13] Material advances in textile and steel production, mining, weaponry, manufacture, and transportation were testament to humans' ability to harness steam and electric power, proffering benefits and pleasures to all mankind. Material progress in the West was so powerful that it dwarfed Chinese civilization of the last two thousand years. Wedded with the Enlightenment, scientific discourse had social implications: It gave rise to intellectual transformation and social revolutions across the European continent and America. The social changes dismantled the "aristocracy and the privileged, granting equality and suffrage to common people. People's hearts brimmed with concepts of liberty, equality, and social democracy."[14]

But science and technology has brought progress as well as regression. Despite all modernity's accomplishments and advances, Lu Xun remained deeply wary of the tendency of modern civilization to slip back to a dark age. Mesmerized by triumphant technology, industrialization, and material progress, and dazzled by benefits of a material life, the elite and the public began to worship science and technology. Science and technology took on a new mythical power and became a new faith. The twin was universally accepted as "the ultimate criteria, held up as the foundation for understanding and judging all things related to the realm of spirit" and "became the repository of all values, the sole vessel of esteem and worship."[15] On the other hand, democratization and decentering, rather than making modern society more liberal and equal, degenerated into a "tyranny of the majority" that encouraged conformity and homogeneity by suppressing individuality and autonomy. Seeing how science became a new gospel that overrides political, ethical, and social aspects of a human civilization, Lu Xun raised serious questions about the imbalances of modernity.

In the essay "Moluo shili shuo" 摩羅詩力說 (On the Power of Mara Poetry), Lu Xun discerns the seeds of environmental crisis in the fetishism of technology and science and the feverish quest of wealth and power. Science discourse renders the natural world desolate, meaningless, and disenchanted; the nation's developmental agendas set the eye on utilitarian gains and seek to alter nature in order to extract its wealth and resources. The knowledge of science aims at discovering the law of nature in order to dominate and subjugate it. Modernity, buttressed by techno-scientific hegemony, deprives ordinary people of faith, autonomy, and creativity. Enthralled by technological advancements, China's educated elite was chasing superficialities and looking at decayed leaves and branches, totally ignorant of the whole tree.[16]

Appropriated by China's modernizers, the new learning from the West has bred a new knowledge class aligned with the powers that be. Just as the bourgeois elite arose in the aftermath of the Enlightenment and the Scientific Revolution in the West, a Chinese power elite was riding high on the power–knowledge nexus and championed the technological approach to modernity. Bent on seeking wealth and power, this new class was making a deafening noise that Lu Xun dubbed the "malevolent." Targeting this voice in "Po

Esheng lun," Lu Xun strikes at this group for eroding the self-confidence of Chinese tradition and the social fabric of the millennial way of life.

The "hypocritical gentry" comprised mostly the scions of the literati-officials from the Qing dynasty, who donned the mantle of reformers and ideologues in the new age. Faced with China's humiliation at the hand of Western powers and cultural decline, this stratum was very busy and loud in coming up with solutions. Its members talked endlessly about progress, industrialization, national wealth, and power as well as a strong state modeled on the West. They brandished terms such as "science, utilitarianism, and civilization, believing all they champion to be the highest order of correctness and beyond refutation" (Lu Xun 8: 26; Kowallis 48). In command of publicity, press, and public opinion, they lectured to ordinary people, dazzling them with newfangled modern ideas. Having studied in the West, they took upon themselves to import Western customs and fashions, absorb scientific phraseology, and impose Western standards on what they deemed an uncivilized and backward China. Puffed up by cultural capital, trendy rhetoric, and technical clout, they strutted down the street, and greeted Westerners with handshake and smile. Although they put on airs as the best and brightest, they were in fact egoists obsessed with petty gains and self-interest.

The hypocrite gentry attacked folk religions and superstition, following the heels of modernization of the Westernization Movement, which privileged technology and industrialization over cultural and religious traditions. The new educated elite made a point of trashing popular religious practices in remote rural areas. Brandishing the rhetoric of enlightenment and science, they wrote books and articles denouncing superstition as unscientific and irrational. Under the mantle of national salvation and progress, they campaigned to eradicate "superstitious" ideas, prohibited local folk festivals and customs like parades of gods in the harvest times, and destroyed Buddhist temples and ancestral shrines. They built new schools to teach new learning but set themselves up headmasters and authorities. They used a rationalist language to debunk mysteries and faiths associated with traditional mentalities and folklore. Having picked up a smattering of chemistry, they claimed that there is no such a thing as the "will-of-a-wisp." There is nothing mysterious and meaningful about it, they said; it is just an element called "phosphorous" that glows in the dark." Having paged through books of physiology, they talked about the human body in terms of molecule biology, proclaiming that the body is only "composed of cells" and "how can there be a soul?"[17]

To counter this talk, Lu Xun wrote: "We must rid ourselves of the hypocrite gentry. Superstition may remain" 偽士當去，迷信可存 (weishi dangqu, mixin kechun). The Chinese character 偽 for hypocrite suggests that the educated elite was touting the supreme benefits of science and progress without grasping the meaning. They pretended to lay claim to the authority and prestige of science and technology in order to seek status, power, and interest. Cynical, self-serving and in bad faith, they were a bunch of narrow-minded and selfish crooks who deployed high-sounding phrases on public platforms for emolument and gains. They "live by wine and food alone" (Lu Xun 8: 29; Kowallis 52).

"Superstition may remain." Coming out of Lu Xun, the militant critic against Chinese tradition and the "standard bearer" of the New Culture in Mao Zedong's words, the remark is paradoxical. Lu Xun's relentless critique of Confucian dogma and popular superstition is beyond dispute. Wang Hui, however, has noted a streak of appreciation in Lu Xun's thinking about super-stition and religion. In "superstition," Lu Xun sees an enduring metaphysical need and a desire among ordinary Chinese to transcend the limits of every-day life and work.[18] This metaphysical need, I would argue, also suggests an expression of the ecological sensibility rooted in the cosmological resonance between humans, heaven, and earth. As we saw earlier, Lu Xun asserts that Chinese have had a "reverence for heaven and earth" through their long his-tory, and this mindset "extends to countless host of material things." The peasants held that "plants and rocks have "mysterious and supernatural prop-erties" rather than mere objects subject to measurement and used for utilitar-ian purposes. Thanks to this sensibility and metaphysical yearning, natural objects were viewed as different from ordinary objects." Seeing in objects as enchanted with religious significance, Lu Xun goes on:

> those who are not satisfied by material life alone will inevitably have spiritual needs. . . . Although our men of ambition in Chin regard this as superstition, I would say that such things [mythology and religion] are indicative of the desires of a people who sought to improve themselves by means of transcending a wholly relative and limiting reality in order to enter the lofty realm of unlimited absoluteness. The human heart requires something to fall back on; without some form of faith, the wherewithal to endure is absent, hence the creation of religions was inevitable. (Lu Xun 8: 27; Kowallis 46)

> 倘其不安物質之生活，則自必有形上之需求…雖中國志士謂之迷，而吾則謂此乃向上之民，欲離是有限相對之現世，以趣無限絕對之至上者也。人心必有所憑依，非信無以立，宗教之作，不可已矣

In Wang Hui's analysis, Lu Xun's defense of superstition affirms the enduring validity of religion, but the concept of religion has little to do with religious institutions, ecclesiastic hierarchy, or dogmatism. Rather, by *mixin*, Lu Xun is searching for a genuine faith in a world of chicanery, duplicity, and hypocrisy. He is searching some forms of hope and spirituality within mate-rial survival. Cherished and practiced by peasants and folks, faith persists in modern times. Rooted in ancient Chinese cosmology, the faith in the organic link of humans and nature is kept alive in long-entrenched grassroots practices, and "is to be found only in the written accounts of the ancients" and "among peasants who still preserve the ways of life of their ancestors." As Wang Hui notes, "the only people of true faith are the common people living in the countryside and ordinary peasants, as opposed to the hypocritical gentry."[19]

Faith persists in modern times because it could constantly renew the organic ties between humans, earth, and heaven, re-enchanting aura and meaning among all things. Faith functions to fulfill spiritual needs and forge ethical and social relations. Rather than false consciousness and self-delusion, "superstition" in Lu Xun affirms such a faith and entails a belief in an ecological, integrated interrelation of nature, culture, language, and imagination. It is from the vantage of this organic concord of earth, humans, and heaven that the malevolent voice sounds jarring, evil, and discordant.

Frankfurt School critical ecology posits an enchanted picture of nature that is redolent with aura and memory. Such nature is a sign of substantive rationality as opposed to instrumental rationality. Murray Bookchin, the pioneer in social ecology inspired by the Frankfurt School, explained that substantive rationality organizes and motivates humans and nature as an immanent force and imparts "meaning and coherence to reality at all levels,"[20] envisioning a meaningful universe resonant with human fate. In contrast, instrumental rationality sees human activity on planet Earth in terms of exploitation and domination of nature as resources for wealth and profit. Based on this distinction, Walter Benjamin distinguished between the ancient stones consonant with human beings and useful rocks earmarked for techno-industrial extraction. Premised on an organic human–nature relationship, Benjamin's charmed stones run a conversation—an animistic language of care, sympathy, and *correspondance*—with mankind before they were stripped of their aura and intimacy by instrumental rationality and techno-industrial agendas. This organic aura derives from a cosmology. The story "The Alexandrite" written by Leskov well illustrates this cosmology in a stone. In it, the gem engraver Wenzel has created the chrysoberyl, a precious stone. Discerning the connection between a handcrafted gem and the cosmos, Benjamin sees Leskov the storyteller as someone endowed with a cosmological imagination. He is not only able to venture "into the depths of inanimate nature" but also transcend earthly existence to rise to the celestial height.[21] This capacity is integral to Benjamin's notion of *correspondance*. Derived from Baudelaire and key to the theory of aura, *correspondance* describes human responsiveness to and intimacy with natural objects and verse versa: "To perceive the aura of an object we look at means to invest it with the ability look at us in return."[22] *Correspondance* cuts across centuries and forges a resonant, secret agreement with prehistory. In the disenchanted world of urbanity, industrialization, and spiritual depletion, however, Baudelaire's poems present an attempt to retrieve memories that correspond with the "data of prehistory," so that the "murmur of the past may be heard in the correspondences." This transhistorical "secret agreement" is the canonical experience of substantive rationality, and its original home is to be found "in a previous life."[23] Richard Wolin speaks of *correspondance* as an "animistic relation to nature." Nature "ceases to be viewed as mere fodder for technical exploitation and is instead regarded as itself ensouled." The experience of *correspondance* "harks back to an unhistorical state of reconciliation, a state prior to that point where the species had succeeded in individualizing itself vis-à-vis primordial nature, in which man and nature existed in a condition of immediate, undifferentiated unity."[24]

The techno-scientific worldview strips the aura of correspondence from humans and nature. Lu Xun notes a scientific epistemology at work in the technocratic language of the hypocrite gentry. In their dismissal of superstition, they claimed that the human body is composed of cells and that "will-o'-the-wisp" is but a phenomenon made of phosphorous elements. But mysteries of nature, counters Lu Xun, cannot be explained away by "a primer on the natural sciences" (Lu Xun 8: 28; Kowallis 51). In the rise of Hinduism and Judaism, Lu Xun saw the religious outlook of nature consonant with human culture and recognized the human needs and desire to apprehend terrifying natural phenomena and to assuage fear. Religious attitudes toward nature "are indicative of the desire of a people who sought to improve themselves by transcending a wholly relative and limiting reality in order to enter the realm of unlimited absolutes" (Lu Xun 8: 27; Kowallis 49). "In the realm of everyday life and production, people need festivals, custom, ritual, and religious practice. The peasants toil all year around and the festival offers a break whereby they give thanks to the gods and indulge themselves." Through sacrificial rituals, the peasants "replenish themselves spiritually and physically" (Lu Xun 8: 29; Kowallis 53).

The abyss between secular enlightenment and true belief is what makes the educated elite hypocritical and pretentious. True belief (*zhenxin* 真信), inherent in the organic relationship between humans and nature as well as the yearning for transcendence, finds expression in the genuine voice of the heart. The belief and the voice of the heart constitute "the fabric of society," the nation, and "the alters of our ancestors," which are imperiled by the hypocrite gentry (Lu Xun 8: 28; Kowallis 51).

Lu Xun pits the voice of the heart (*xinsheng* 心声) against the malevolent noise. The voice of the heart is the expression springing up from the inner heart and is premised on a continuity of inner and outer nature. The veneration of nature not only serves religious and spiritual purposes but also alludes to the underlying interdependence between man and nature. As Lu Xun observes, "human beings, in their observation of phenomenon and in their investigation of the material world, felt as though everything embodied some mysterious properties from which we derive everything that is beautiful and ingenious" (Lu Xun 8: 30; Kowallis 50). The beautiful is more than a matter of aesthetics; it entails a religious view of poetry and the world: Poetry is rooted in nature and human–nature bonds are expressible through the transparent medium of human voice. On the other hand, human resonance with nature provides an exhaustible source for poetry, imagination, and myths. This "aesthetic-religious view," based on the embrace of heaven and earth, entails a deep ecological intimacy between humans and nature. The true voice of the heart arises from the inner light (*neiyao* 內耀) and promises to deliver China from the world of chicanery and hypocrisy. Miming the voice of nature, the voice of the heart sounds "like a clap of spring thunder and bestirs the hundreds plants to bud; like a crack of dawn in the east, it heralds "the passing of night" (Lu Xun 8: 23; Kowallis 40).

In "Moluo shi lishuo," Lu Xun links the voice of the heart to primordial strata and the robust nature of premodern people, folk culture, and native sentiments. Citing a case of atavism in biology, he accounts for the reappearance of the primitive in the new, as traces of wild zebras pop up as irruptions of untamed, primordial nature amidst domesticated animals. This bio-evolutional event demonstrates indomitable raw nature bursting through culture.[25] This irruption offers a glimpse of the gap between nature and culture. Poetry is to close this gap. The unity between voice and nature is the key to vital energy that jumpstarts and re-animates the life of a civilization. In this eco-mimetic vein, civilization grows like a tree or a natural process, and the voice of the heart constitutes its most powerful written legacy. Primitive humans were connected with the mysteries of nature and deeply intertwined with mythological imaginations. Chinese ancestors had achieved an innately spiritual communion with nature, which, when verbalized and expressed, gave rise to songs and poems.

The voice-nature continuum allows Lu Xun to discern ecological elements in the *Mara* poets. English poet Percy Bysshe Shelley, alienated by the rigidly moralistic Victorian England, placed a deep trust in nature. His immersion in wilderness for poetic inspiration contrasts sharply with the behavior of the hypocrite gentry of his society, who eyed nature as economic resources. Like the Chinese hypocrites, the British bourgeoisie pretended to be the vanguards of progress and promoted a techno-scientific agenda, seeking to control nature and discover its laws in pursuit of wealth and power. The bourgeois elite was immune to the inner life of nature and tone-deaf to its poetry. By contrast, Shelley, possessing a pure and noble imagination, roamed nature, and felt its secret recesses. Everything that meets his eyes seems to have affection and mutual sympathy. Setting his heartstrings in tune with the rhythm of nature, Shelley created a unique and unparalleled body of poetry.[26] Russian writer Lermontov was another example of empathy and dialogue between humans and nature, figuring as a polar opposite of Pushkin. Pushkin advocated a beastly patriotism 獸性愛國 (*shouxing aiguo*), a term Lu Xun coined to attack chauvinists in Russia and China for their imperialist agenda to colonize other regions and plunder resources of other lands. Lermontov was patriotic, but he loved the soil and landscape of his native land and was deeply in touch with nature and folk culture. "All his truest love was for the villages and the steppes, and he was deeply attached to the native folks and village life and customs, especially the natives of Caucasus, a region that had risen in arms against Russia to achieve its own freedom."[27]

CRITIQUES OF THE HYPOCRITE GENTRY IN LU XUN'S SHORT STORIES

Two stories in Lu Xun's collection *Gushi xinbian* 故事新編 (Old Tales Retold) address mythology, cosmology, and human–nature resonance. Instead of nostalgia, Lu Xun invoked archaic elements in a critical gesture to shed light on the ways in which a civilization dominated by the hypocrite gentry trails a record of domination of nature and people. Ecological problems could be traced to problems of social and political ecology. "Butian" 補天 (Mending Heaven) tells the story of the Goddess Nü Wa 女媧. Nü Wa comes on the scene as the progenitor of a self-generating universe. A robust body indistinguishable from mountains, sky, and oceans, she wakes up from a dream to wade into a spring-like, thriving universe. An embodiment of primordial forces, her figure, tinged with pink color, envelops and washes over heaven and earth. "She walked though this flesh-pink universe to the sea, and the lines of her body merged with the luminous, rose-tinted ocean, only a zone of pure white remaining visible at her waist."[28] The difficulty of defining her in a human shape points to the all-encompassing primordial force and libidinal energy that constitute the throbbing life of the cosmos. In Chinese cosmology, there is no such entity as a transcendent God that stands outside the created realm. Tu Weiming has argued that the Chinese creation myth, distinct from the Cartesian dichotomy of spirit and matter, envisions a process of self-creation and self-generation, enmeshing "spirit and matter as an undifferentiated whole."[29] Lu Xun's Nü Wa fits this description. Both progenitor and contour of heaven and earth, she uses wisteria to make adjustment to nature while creating little human beings on the ground.

Before long, heaven splits asunder and the earth breaks apart, leading to a scene of utter chaos and of environmental collapse. "The earth was heaving and shaking" and "water, sand and rocks were raining down on her from behind" (Lu Xun 2: 345; Yang 8). Nü Wa comes to the rescue and works hard to mend the broken cosmos, like a peasant toiling in the field. "With the wrinkled brows she looked around and reflected before wringing the water out of her hair. She threw back her shoulders, starting with fresh energy to gather reeds" and build fires to burn stones for mending heaven (Lu Xun 2: 351; Yang 11). The small creatures she has created show up, and they appear to have turned into a different set. Some demand of her the gift of immortality and others, seeing her naked, accuse her of immoral exposure. These petitioners are unmistakably portrayed as the hypocrite gentry. One petitioner, a minister from the imperial court, is all pomp and circumstance. Hung with thick folds of drapery from head to foot, he wears a dozen supernumerary ribbons at this waist. Crowned by a small black oblong board (Lu Xun 2: 351; Yang 12), he presents "a highly polished green bamboo tablet on which were two columns of minute black specks" (Lu Xun 2: 352; Yang 13). The craftsmanship of graven letters is impressive and complex yet incomprehensible to Nü Wa. The official charges her for being lewd and condemns her nakedness as "immoral, an offense against etiquette and a breach against rules and conduct fit for beasts" and "forbidden by the laws of the land." Taken aback by such accusations, Nü Wa feels that "she could have no real communication with these creatures" (Lu Xun 2: 353; Yang 13). Nü Wa encounters another incarnation of the hypocrite, this time a bunch of soldiers at war. In the killing field, the soldiers led by the giant Gong Gong 共工 complain that their king has knocked himself against Mount *Buzhou* 不周山 and has broken this pillar of heaven, and that their defeat signals the injustice of Heaven. Their enemy on the other side, represented by the mythical figure Zhuan Xu 颛顼—the legendary descendant of Yellow Emperor, claims that Gong Gong harbors evil ambition and his defeat is testament to heavenly justice. The appeal to heavenly justice in the middle of a brutal war is open to two readings. While Nü Wa is mending Heaven's rupture, the rupture among humans and nonhuman creatures seems incomprehensible to her. The high-sounding appeal to heavenly justice by the warring armies points to another trait of the hypocrite gentry, recalling Lu Xun's satirical jibe at cats.

In his essay, "Gou, mao, shu" 狗·猫·鼠 (Dogs, cats, and mice) Lu Xun suggests that although no sharp distinction can be drawn between humans and animals, animals appear more genuine and honest than humans. "Animals act according to their nature, and whether right or wrong never try to justify their actions."[30] "Although the animal kingdom is by no means as free and easy as the ancients imagined, here is less tiresome shamming there than in the world of men" (233). "Shamming" is the rendering of *lusu zuozuo* 嚕嗉做作 by Yang Hsien-yi and Gladys Yang. *Lusu* implies verbiage and incessant talk of rationalization, and *zuozuo* refers to performative, affected gestures to grab attention. The phrase captures the hypocritical behavior of the educated elite. Vultures and beasts, Lu Xun goes on, may prey on weak creatures, but they have "never hoisted the banners of 'justice' and 'right' to make their victim admire and praise them right to the time they are devoured" (233). By contrast, cats take to more shamming or indulge in the habit of performative self-display and self-justification, as they caterwaul before they get down to the real act of mating. Their pretentious showiness characterizes the rich and powerful in China as they constantly display pomp and circumstance. In a private wedding ceremony, for example, the marrying families would stage a huge show, complete with paraphernalia of rites and ritual and noisy celebratory rhetoric before sexual union.[31] A counterpart to the hypocrite gentry, the cats are full of affectation and empty talk *konghua* 空話. In "Butian," the hypocrites, be they officials and warriors, are howling creatures constantly making loud and shrieking noises in elegant and melodramatic fashion—malevolent voices 惡聲 totally incomprehensible to Nü Wa.

Though a cosmic creator and fixer, Nü Wa bears human characteristics and weaknesses. She must work hard in her repair work and consequently dies of exhaustion. The victorious army of Zhuan Xu now comes to settle on her belly, the most fertile portion of her body. Taking up the prime spot for extracting natural resources, the new regime alters the words on their standard into "Entrails of Nü Wa." The new dynasty claims the Goddess as the origin of its illustrious genealogy, which bequeaths to a stream of dynasties as her "true descendants." This ending suggests that the attempt to attain immortality through aligning human life with the Goddess will come to nothing. Not only nature decays and dies but also the giant tortoises, the ferry to the Fairy Mount sent by Nü Wa, has been stranded at the outset. Although emperor after emperor has sought the ways to reach the Iles of Fairies, they are merely chasing an impossible dream.

The story "Lishui" 理水 (Curbing the Flood) invokes the ancient myth of *Da Yu zhishui* 大禹治水. In containing the flood, Da Yu spent years during an epic campaign to channel the flood without seeing his family. Lu Xun's retelling of the legend is concerned with environmental calamities and human agency to alter nature in order to protect communities from harm's way. Yet similar to "Butian," the story depicts a social structure dominated by the hypocrite gentry and bureaucratic hierarchy. The officials, scientists, and scholars command all resources, knowledge, and power to fight the flood, yet they stand as barrier to curbing the flood and helping the people. Indeed, the social ecology, dominated by a modern technocratic and cultural elite, aggravates natural disaster and increases the misery of disaster victims.

Interweaving scientific jargon, technical gadgets, cultural capital, and food science with myths and imperial bureaucracy, the story reads like a work of science fiction. It mounts a sharp critique of the hypocrite gentry across the

divide of modernity and antiquity. In the opening scene, a huge flood is rag-
ing, "encircling mountains and engulfing hills" (Lu Xun 2: 371; Yang 30).
The privileged subjects of Emperor Shun flock to the heights but the victims
or subalterns (*xiamin* 下民) are reduced to "environmental refugees," hanging on to tree tops and taking to fragile rafts. While the empire is partially
shut down, scholars and scientists—alluding to a group of nonchalant scholars unconcerned with the Japanese invasion of China in the 1930s—find refuge in the Mount of Culture 文化山 (*wenhua shan*), an island of a scientific
and intellectual academy surrounded by flood and misery. They are having
great fun and continue to conduct exoteric research and scientific investigation on the Mount. Food is brought in by "flying chariot from the Kingdom
of Marvelous Artisans" and the deliverers communicate with scholars in
English. Knowing that the government has sent Da Yu to contain the flood,
scholars argue back and forth on whether Yu is a fish or a man. They deploy
positivist epistemology and spend long hours in hairsplitting etymological
and genealogical research, only to find that underprivileged subalterns have
survived for generations without keeping a genealogy. To prove the non-
existence of Yu, a scholar uses a paste of breadcrumbs mixed with water
and charcoal to write on the trees in minute tadpoles-shaped characters. The
esoteric, meticulous craftsmanship of writing, evocative of the unreadable
Chinese scripts for Nü Wa, becomes an exhibit, a spectacle of modern cultural industry. The poor visitors have to pay with elm leaves or duckweed to
admire these high-minded, elegant heritages.

Although Yu is a real official charged by Emperor Shun with curbing the
flood, the official commissioners come first to the disaster zone to inspect
damages. Two middle-aged, corpulent officials arrive with a huge entourage
and an array of armed guards. As they enquire about food shortage in disaster
areas, the answer from the cultural and scientific elite is "things are ok." The
flying chariot delivers food every few weeks. Although the subalterns feed
on leaves and weeds, one expert, schooled on *Shennong bencao* 神農本草
(Materia Medica of Emperor Shennon) talks glibly like a nutritionist about
the health benefits weeds eating: Vitamin W in elm leaves and iodine in
seaweed. With scientific talk glossing over the disaster and suffering, the
bureaucrats begin to enjoy their sojourn in the Mount of Culture. They
indulge in the poetic pleasure of natural beauty: admiring an umbrella-shaped
old pine on the highest peak and going fishing for eels behind the mountains.
On Day Five, the officials summon a representative from the disaster victims
for information. Initially, scared and daunted by official entourage and for-
bidding guards, this guy, when asked about food shortage, answers that the
masses can get by and that their meager eating is quite sustainable for a long
haul. Following the same hypocritical trick of the scholars, he touts the power
of language to turn low into high, and ugly into beautiful: The subalterns

have given poetic and elegant labels to their wretched food. "Water-weed" 水苔 (*shuitai*) is deliciously named "Slippery Emerald Soup" 滑溜翡翠湯 (*hualiu feicui tang*), and elm leaves are graced with "First-at-Court Gruel" (一品當朝羹 *yinpin dangchao geng*) (Lu Xun 2: 378; Yang 40). In the face of calamities and starvation, officials and subalterns, high and low alike play a game of shamming and indulge in pompous talk, scientific phraseology, and aesthetic taste.

The final report of flood is circulated at the court banquet. In chatting about the disaster areas they just left behind, the officials and guests talk about how muddied floodwater gleams like gold, full of succulent eels and slippery duckweed. The technocratic talk, appraisal of natural beauty, and elegant phrases are nothing but the malevolent voice of the hypocrite gentry. These bureaucrats and leaders are riding high on a civilization that is teetering on the brink of collapse.

Then, Da Yu, the flood manager, comes on the scene, demonstrating a real legend and proactive human spirit. Unlike the scholars and officials who thrive on empty talk and shamming, Yu has his feet on the ground. Though a high official, he dresses and behaves like a peasant: cutting the figure of a thin hulk of a man with huge hands and feet (Lu Xun 2: 380; Yang 42). He has informal and unceremonious manners, wears no socks, and "the soles of his feet were covered with calluses the size of chestnuts." He goes around without regal insignia and is always followed by a "band of beggarly looking followers" (Lu Xun 2: 385; Yang 48). Cutting to the chase amid the endless talk of exotic food and natural beauty, he decides to use the method of channeling instead of damming—his father's method, to control the flood. This causes another vacuous debate among high officials. Several charge him for his violation of filial piety in breaking with his father. But his adoption of the new strategy does not come easy. It arises from Da Yu's painstaking investigation of the flood conditions all over the land and by gathering wisdom and advice from the flood victims.

After Yu has successfully curbed the flood and re-stored social order, Emperor Shun asks for the secret of his success. His answer is "Keep hard at it everyday." Coming from an honest, upright, virtuous man, the answer embodies what Lu Xun deems the voice of the heart and true belief—signs of exemplary virtue. Yu's success represents a commitment to the wellbeing of all under heaven and a grounded practice to make human survival sustainable in the natural environment. Traveling up and down the empire through water and on land, on foot and by carriage, Yu exemplifies active human engagement with nature, serving as a medium between social ecology and natural ecology. He witnesses people being swallowed up in the water. He fells trees and channels flood water in the fields into the rivers, and the water of the rivers into the sea. He makes sure that everyone "had rice and meat to eat." He distributes resources and relocates people to better locales. "So at last everyone has settled down in peace everywhere and order reigns" (Lu Xun 2: 385; Yang 49).

In the wake of Yu's success, a prosperous society and thriving markets emerge. Honoring his great deeds, can-do spirit, and commitment to the common good, Emperor Shun made Yu a model for the population to emulate. The imperial decree is fraught with further perils. By intensifying visual and performative spectacle of exemplariness, the program re-packages virtuous substance and heroic conduct into trappings of cultural capital for visual and popular display and consumption. On returning to the capital city, Yu begins to lose touch and is being folded back quickly into the courtly life as usual, finding a privileged spot in the ranks of hypocrites and bureaucrats. "Though he dressed simply at home, when it came to sacrifices or public occasions he made a great display" (Lu Xun 2: 386; Yang 50). In court proceedings, "he put on splendid robes." In such imperial phantasmagoria, outer nature and inner nature are alienated and deployed as trappings in shamming procedures. "Then such peace reigned throughout the world that even wild beasts danced and phoenixes flew down to join in the fun" (Lu Xun 2: 386; Yang 50).

Like Frankfurt school critics, Lu Xun mounts a sharp critique of the modern myth of progress, science, and technology. By evoking ancient myths, cosmology, and ecological resonance of humans with nature, Lu Xun lampoons the hypocrisy and illusoriness of the hypocrite gentry. This privileged and learned class includes China's reformers and westernizers at the turn of the twentieth century, who were in thrall to technology and science and obsessed with the pursuit of wealth and power. In Lu Xun's stories, the knowledge elite in modern China rub shoulders with bureaucrats and literati of the imperial dynasties. By indulging cynically in empty talk and performing vacuous rituals, by wielding the rhetoric of science and rationality, this elite set valorizes the myth of modernity by trashing "superstition. Lu Xun's defense of superstition articulates the millennial faith in the ecological interconnection of humans and nature, and expresses the intimate connection between faith, voice, and poetic imagination. In retelling the legend of Da Yu, Lu Xun depicts the organic reciprocity of humans and nature and proactive human engagement with environmental crises.

NOTES

1. Lu Xun, "Po ersheng lun," *Lu Xun quanji* [Complete works of Lu Xun], vol. 8 (Beijing: Renmin wenxue, 1981), 28. I use Jon Kowallis' English translation of "Toward a Refutation of Malevolent Voices," *boundary 2* 38.2 (2011): 51. Further references to this essay will be in parentheses with two page numbers.

2. Walter Benjamin, *Illuminations* (New York: Schocken Book, 1968), 96.

3. Benjamin 236.

4. Benjamin 223.

5. Benjamin 242.

6. Max Horkheimer and Theodor Adorno, *Dialectic of Enlightenment*, ed. Gunzelin S. Noerr and tr. Edmund Jephcott (Stanford, CA: Stanford University Press, 2002).

7. Benjamin 156–157.

8. Benjamin 156.

9. Cited in Richard Wolin, *Walter Benjamin: An Aesthetic of Redemption* (Berkeley and Los Angeles: University of California Press, 1994), 234.

10. *Correspondance* is a French word in Baudelaire and theorized by Benjamin to denote mythical aura and resonance between humans and nature.

11. *Lu Xun quanji*, vol. 1 (Beijing: Renmin wenxue, 1981), 415.

12. Lu Xun, *Luxun quanji*, vol. 1, 25.

13. Lu Xun, *Lu Xun quanji*, vol. 1, 47. In translating Lu Xun's essay into English, I use Jon Kowallis' excellent but unpublished translation and add my modifications. My gratitude to Kowallis for sharing his translation with me.

14. *Lu Xun quanji*, vol. 1, 48.

15. *Lu Xun quanji*, vol. 1, 48.

16. *Lu Xun quanji*, vol. 1. 86.

17. *Lu Xun quanji*, vol. 8, 29.

18. Wang Hui, "The Voices of Good and Evil: What is Enlightenment: Re-reading Lu Xun's on the Refutation of Malevolent Voices," *boundary 2* 38.2 (2011): 67–123.

19. Wang Hui 113.

20. Murray Bookchin, *The Ecology of Freedom: The Emergence and Dissolution of Hierarchy* (Palo Alto, CA: Cheshire Books, 1982), 10.

21. Benjamin, *Illuminations* 107.

22. Benjamin 188.

23. Benjamin 182.

24. Wolin 236.

25. *Lu Xun quanji*, vol. 1, 64.

26. *Lu Xun quanji*, vol. 1, 86.

27. *Lu Xun quanji*, vol. 1, 91.

28. *Lu Xun quanji*, vol. 2, 345–346. I use Yang Hsien-yi and Gladys Yang's translation in *Old Tales Retold* (Beijing: Foreign Language Press, 1972), 6. Further references to "Mending Heaven" and Curbing the Flood" will be in parentheses with two page numbers from *Lu Xun quanji*, vol. 2 and Yang's translation.

29. Tu Weiming, "The Continuity of Being: Chinese Visions of Nature," in *Chinese Aesthetics and Literature*, ed. Corrine Dale (Albany: State University of New York Press, 2004), 29.

30. *Lu Xun quanji*, vol. 2, 233. I use Yang Hsien-yi and Gladys Yang's English translation of Lu Xun collection of memory pieces in Lu Hsun, *Dawn Blossoms Plucked at Dusk* (Peking: Foreign Languages Press, 1976). References to Lu Xun's essay will be in parentheses in the text.

31. *Lu Xun quanji*, vol. 2, 233.

WORKS CITED

Benjamin, Walter. *Illuminations*. New York: Schocken Book, 1968.

Bookchin, Murray. *The Ecology of Freedom: The Emergence and Dissolution of Hierarchy*. Palo Alto, CA: Cheshire Books, 1982.

Horkheimer, Max and Theodor Adorno. *Dialectic of Enlightenment*. Edited by Gunzelin S. Noerr and translated by Edmund Jephcott. Stanford, CA: Stanford University Press, 2002.

Kowallis, Jon E. "Toward a Refutation of Malevolent Voices." *boundary 2* 38.2 (2011): 39–62.

Lu, Hsun. *Dawn Blossoms Plucked at Dusk*. Translated by Yang Hsien-yi and Gladys Yang. Peking: Foreign Languages Press, 1976.

Lu, Xun. *Old Tales Retold*. Translated by Yang Hsien-yi and Gladys Yang. Beijing: Foreign Language Press, 1972.

———. *Lu Xun quanji* 鲁迅全集 [Complete works of Lu Xun]. Vols. 1, 2 and 8. Beijing: Renmin wenxue, 1981.

Tu, Weiming, "The Continuity of Being: Chinese Visions of Nature." In *Chinese Aesthetics and Literature*, edited by Corrine Dale. Albany: State University of New York Press, 2004, 27–40.

Wang, Hui, "The Voices of Good and Evil: What is Enlightenment: ReReading Lu Xun's On the Refutation of Malevolent Voices." *boundary 2* 38.2 (2011): 67–123.

Wolin, Richard. *Walter Benjamin: An Aesthetic of Redemption*. Berkeley and Los Angeles: University of California Press, 1994.

Chapter 3

Planetary Healing through the Ecological Equilibrium of *Ziran*

A Daoist Therapy for the Anthropocene

Jialuan Li and Qingqi Wei

THE HISTORICAL CONTEXT OF DAOISM IN THE ANTHROPOCENE

The historical contextual relevance of Daoism to the modern ecological crisis remains inadequately studied despite the abundant textual analyses of Daoist canon. Existing studies of Daoism from an ecological perspective have been portraying Daoism as an environmentalist prophecy from a good old time when humans lived in harmony with nature, but the early days of Laozi 老子 (ca. 571–471 BCE), Liezi 列子 (ca. 450–375 BCE), Zhuangzi 庄子 (ca. 369–286 BCE or 275 BCE), and other early Daoist masters, were far from an ecological paradise. The Spring and Autumn period and the Warring States period (770–221 BCE) were a time of human tragedies and ecological disasters, including wars, starvation and social turmoil; and rapid expansion of human settlements and vast invasion of nature. From a sociobiological point of view, the very fact of social chaos and human habitat expansion attests to the ecological tension. Humans fell into the Malthusian Trap, and the human population reached a threshold beyond the carrying capacity of the environment.

The Anthropocene, a term often attributed to Nobel Laureate Paul J. Crutzen (1933–) and Eugene Stoermer (1934–2012), refers to a proposed epoch of significant human impacts on Earth's geology and ecosystems, including but not limited to anthropogenic climate change. Humans have exerted so much influence upon the planet that they as a species collectively have become a geological force, changing temporarily or permanently the biosphere, the atmosphere and even the lithosphere. In 2000, Paul Crutzen,

IGBP[1] Vice-chair at the time, and Eugene F. Stoermer proposed in IGBP's Global Change newsletter 41 that humanity had driven the world into a new geological epoch, "the Anthropocene." Two years later, Crutzen published his famous article "Geology of Mankind" in the journal *Nature*, gaining much attention within the scientific community. In this article, he listed the scientific markers of anthropogenic influence on the Earth geology, traced the history of studies on mankind's growing influence on the environment, and cited similar terms that predated "the Anthropocene," such as the Italian geologist Antonio Stoppani's "anthropozoic era" and "noösphere"—the world of thought—coined by the Russian geologist V. I. Vernadsky, the French Jesuit P. Teilhard de Chardin, and E. Le Roy in 1924.

But the specific starting point of the Anthropocene remains contentious. Crutzen proposed in "Geology of Mankind" that "the Anthropocene could be said to have started in the late eighteenth century, when analyses of air trapped in polar ice showed the beginning of growing global concentrations of carbon dioxide and methane" (23). Such a starting date coincides with the Industrial Revolution marked by James Watt's improvement or (re)invention of the steam engine in 1784. The palaeoclimatologist William Ruddiman proposed the Early Anthropocene Hypothesis, which posits that "the Anthropocene actually began thousands of years ago as a result of the discovery of agriculture and subsequent technological innovations in the practice of farming" (261). This proposal of the Agricultural Revolution (also known as the Neolithic Revolution) some 8,000 to 10,000 years ago as the starting point of the Anthropocene is the most ancient among the proposed dates. In 2009, the International Commission on Stratigraphy (ICS) established the International Anthropocene Working Group to investigate whether the Anthropocene should be adopted as part of the geologic time scale, and in January 2015, 26 of the 38 members of the International Anthropocene Working Group voted to suggest the Trinity nuclear test on July 16, 1945, as the starting point of the Anthropocene. The year of 1945 or mid-twentieth century has been termed as the Great Acceleration, when the severity of human impacts on the Earth geology and ecosystems significantly increased. Now it is up to the ICS's approval and the final ratification of the International Union of Geological Sciences (IUGS) before the Anthropocene is officially adopted as part of the geologic time scale and a starting point is assigned to this new epoch. Although the term has not yet been officially recognized as a subdivision of geological time, one fact stands undeniable: the human kind has significantly changed the ecology and incurred all sorts of global and local disasters.

To work around the ongoing contention for the starting point of the Anthropocene, I would propose that the Agricultural Revolution signified the beginning of "the minor Anthropocene," a time when the human species still had relatively acceptable impacts on the planet; while the Industrial

Revolution or the 1945 Trinity nuclear test—the currently favored starting point of the Anthropocene—marked the start of "the severe Anthropocene," a time when humans have devastating impacts on Earth. This distinction is self-evident if we see the accelerating degree of influence humans have been exerting on the Earth ecology and geology. Apparently the term "the minor Anthropocene" works best if the Agricultural Revolution is accepted as the beginning of the Anthropocene. Should the ICS and the IUGS decide to use the 1945 Trinity nuclear test as the formal starting point of the Anthropocene, "the minor Anthropocene" can still be rightly used to refer to the precursory period before "the (severe) Anthropocene" when humans have started exerting influence on the Earth geology and ecology.

If we accept this idea of "the minor Anthropocene" and "the severe Anthropocene," Daoism can be put in a relative position in the long and ongoing history of the Anthropocene. Daoism as a philosophy and as a religion was founded in "the minor Anthropocene," and the bulk of its over two thousand years' history was also in this period. Much like the population decline of any wild species when its number has exceeded the carrying capacity of the environment, humans have constantly been caught in the Malthusian Trap whereby excessive population would cause food shortage, starvation, social turmoil and even wars, which would in return force the population back to an equilibrium with the environment. In his famous *An Essay on the Principle of Population*, Thomas Malthus (1766–1834) established a correlation between population size and food supplies and between population size and national well-being: "The happiness of a country does not depend, absolutely, upon its poverty, or its riches, upon its youth, or its age, upon its being thinly, or fully inhabited, but upon the rapidity with which it is increasing, upon the degree in which the yearly increase of food approaches to the yearly increase of an unrestricted population" (43). When population growth is potentially exponential while the growth of food and other resources is linear, a Malthusian catastrophe of diseases, starvation and wars would be inevitable, decreasing the population to a lower, more sustainable level. Therefore, the American ecologist and philosopher Garrett Hardin (1915–2003), in his famous article "the Tragedy of the Commons," argued that "a finite world can support only a finite population; therefore, population growth must eventually equal zero" (1243) and that "the population problem has no technical solution; it requires a fundamental extension in morality" (1243).

Now, Daoism offers such an "extension in morality" for the Malthusian Trap and other ecological problems. Laozi, Liezi, Zhuangzi, and other early Daoist masters lived in a period of social transition; the effects of "the minor Anthropocene" were apparent in the forms of chaotic wars between rivaling states. Zhu Kezhen (1890–1974) noticed the existence of elephants in northern China, and by comparing archeological evidence and historical records he concluded that "the elephant subfossils unearthed from *Yinxu* must be from local elephants" (17). Mark Elvin made similar observations in *The Retreat of the Elephants: An Environmental History of China*: "Four thousand years ago there were elephants in the area that was later to become Beijing (in the Northeast), and in most of the rest of what was later to be China. Today, the only wild elephants in the People's Republic are those in a few protected enclaves in the Southwest, up against the border with Burma" (9). The retreat of elephants from northern China is just one example of the early symptoms of a changing climate in ancient China. Though without a modern vocabulary, Laozi, Liezi, Zhuangzi, and other early Daoists must have noticed the symptoms of "the minor Anthropocene" by observing the natural phenomena and reading historical records, and they proposed a philosophical and practical turn that could be implemented on the social and personal levels to break out of "the minor Anthropocene" or Anthropocene in general. My key argument is now clear: Daoism was a philosophical reaction to the early symptoms of the Anthropocene, or "the minor Anthropocene," and for this very reason its key ideas can be applied to the Anthropocene, or the severe Anthropocene, as we are undergoing today.

THE TRADITIONAL AND MODERN
INTERPRETATIONS OF *ZIRAN*

The applicability of Daoism to the Anthropocene is more fundamentally based on what it has reflected upon its historical context, what it has taught us on the human-nature relationship. Daoism responded to the social chaos of the time by calling for the abandonment of human ingenuity and the return to nature, which are aptly summarized in the term *ziran*, an ecologically ideal state of equilibrium. The ideal of *ziran* and the methodology of *wuwei*[2] to realize such an ideal are what separate Daoism from other schools of traditional Chinese thought. In Liu Xiaogan's words, "One might even say that the question of whether or not a thinker employs and advocates the ideals of naturalness [*ziran*] and *wuwei* could serve as a standard for determining whether or not the thinker is a Daoist or has been influenced by Daoist ideas" (212). Chen Guying quoted Jiang Xichang in his collection of notes and comments on *Dao De Jing* and discussed the nuanced differences between *ziran* and *wuwei*: "When Laozi mentions the idea of 'naturalness' (*ziran*), he is referring to a state in which not a bit of effort is applied and yet things unfold and develop of their own accord, whereas *wuwei* refers to the idea of following along with the way things naturally are and not adding any human effort" (132). Liu Xiaogan also said that "'naturalness' (*ziran*) is the core value of the thought of Laozi, while *wuwei* is the principle or method for realizing this value in action" (211). *Ziran* serves as a value orientation, an ideal, while *wuwei* falls into the category of methodology. In Liu Xiaogan's words, "Looking at *naturalness* [*ziran*] from the standpoint of value theory and at *wuwei* from the standpoint of methodology, we can avoid the difficulties encountered by approaches that traditionally have taken modern philosophy as their tool and analysis as their primary goal" (215).

The word *ziran* is directly mentioned four times in Laozi's *Dao De Jing*, in Chapters 17, 23, 25, and 51, and there are many other chapters with indirect reference to this core idea. In *Dao De Jing*, *ziran* invariably refers to "naturalness" or "of one's own accord" as *zi* denotes "self" and *ran* means "so." This "self-so" state poses a Daoist ideal: everything—including *Dao*, man, Earth, and Heaven—models themselves after naturalness, or an intrinsic tendency of themselves. Lin Yutang, a well-known Chinese translator, translated *Dao De Jing* into English, and in his translation:

> Tao [*Dao*] gives them birth, Teh (character) [*de*] fosters them. The material world gives them form. The circumstances of the moment complete them. Therefore, all things of the universe worship Tao and exalt Teh. Tao is worshipped and Teh is exalted without anyone's order but is so of its own accord [*ziran*] (Lin, *The Wisdom of Laotse* 242).

Chapter 51 of *Dao De Jing* quoted above starts from *Dao* as the ontological origin of all things and asserts *ziran* as the defining feature of *Dao*. Laozi's *Dao* is both an ontological entity and a prescriptive attribute. As an ontological entity, it gives birth to everything in the Daoist cosmos, so it is very logical for "all things of the universe" to "worship" *Dao* as the highest and ultimate existence. Suppose *Dao* is NOT "of its own accord," then it must inherit its properties from some even higher existence, but there is no higher existence than *Dao* in the entire Daoist cosmos. Therefore, it only makes sense that *Dao* just exists and runs its course in a totally *ziran* manner, dependent on nothing else but itself, or as Laozi put it, "without anyone's order but is so of its own accord" (Lin, *The Wisdom of Laotse* 242). James Miller's recent book *China's Green Religion: Daoism and the Quest for a Sustainable Future* is in line with this interpretation: "There is nothing to which *Dao* itself is subject. Instead, *Dao* is the ground of subjectivity itself: it is the power of things to be autonomous, self-governing and self-creating" (*China's Green Religion* 32). The quality of *ziran* is the true reason why *Dao* is constantly worshipped and why *de* as the manifestation of *Dao* is exalted by everything in the Daoist universe.

> Man models himself after the Earth; The Earth models itself after Heaven; An ancient text reads "man" in place of "King." The Heaven models itself after Tao [*Dao*]; Tao models itself after Nature. (Lin, *The Wisdom of Laotse* 242)

As the key concept of Daoism, *ziran* penetrates into all aspects of human life, including governance, politics, philosophy, and personal self-cultivation. Chapter 25 of *Dao De Jing* quoted above can be seen as a theoretical continuum of Chapter 51 in terms of the idea of *ziran*, extending the discussion into the relationship between humans, Heaven, Earth, *Dao,* and *ziran* itself. Daoism does not take an eco-centric lens upon the universe. Instead, humans exist naturally with nonhuman nature represented by Heaven and Earth, logically including the plants, animals, insects and other beings living between them. It is neither an anthropocentric philosophy/religion where "man is the measure of all things"[3] (Quinton 93), as Protagoras famously claimed, nor is it an ecocentric ecoterrorism that is readily willing to reduce the human population by violence, as deep ecologists might agree with. Protagoras' statement is often interpreted as radical relativism in epistemology, but in the modern context it is also criticized from the ecological perspective. Andrew Linzey and Clair Linzey were outspoken in denying the possibility of Protagoras' statement as a philosophical basis for "an appreciation of the other-than-human": "The Pythagorean maxim 'man is the measure of all things' cannot be a religious starting point [of ethics for animals]" (ii). If given the chance, Laozi and other Daoists might claim something rather different: things are the measure of themselves. In the Daoist cosmos, humans are not the center of things and human standards do not define the value of things.

At any rate, the key point of Chapter 25 lies in the statement on the modeling sequence between "the Great Four"—that is, humans after Earth, and Heaven, *Dao* and *ziran*. As *ziran* functions as a universal value, humans and nonhuman nature (Heaven and the Earth, and all being between them) all model themselves after *ziran*. As the insentient nature naturally follows this principle, the key message Laozi is conveying is that humans should behave according to this universal principle of *ziran* or naturalness, as nonhuman nature does. This goes head on against anthropocentrism, the prime culprit in Western culture that has been pinned down as the primary cause of the ecological crisis. Lynn White Jr. traced the history of Christianity and the Western cultures in general and their functions in forming the modern science and technology. Despite minor mistakes in some details, such as the history of plows, White rightly concluded that "[E]specially in its Western form, Christianity is the most anthropocentric religion the world has seen" (1205). Now Daoism, especially its concept of *ziran*, offers an alternative to Western anthropocentrism. It can serve as an ecological ideal, forming an antithesis of the anthropocentric value system based on conquering, exploitation, and consumption.

In the literal sense of *ziran*, Daoist thinkers were not speaking about nature in its modern sense of the natural environment. Chen Guying said, "Laozi coined the term '*ziran*' and used it in many occasions, but never did he see it as the objective reality of the natural world." (132) But Wei Qingqi has rightly pointed out that "the Chinese word ziran for 'nature' has two apparently discrepant meanings: the original sense of spontaneously being what it is and the modern sense of nature, but this divergence can be perfectly accountable if we presume the modern distinction between nature and nurture." (775) Raymond Williams called nature "the most complex word in the [English] language," and pinpointed its three meanings of this word: "(i) the essential quality and character of something; (ii) the inherent force which directs either the world or human beings or both; (iii) the material world itself, taken as including or not including human beings." (219) Although there is no direct equivalent of "nature (iii)" in the ancient Chinese language and culture, there are several terms that may be approximately translated as "nature (iii)." *Tian* (Heaven/sky), *tiandi* (Heaven and Earth), *tianxia* (all under Heaven), *ye* or *huangye* (wilderness), *shanshui* (mountains and waters, often translated as landscape), when put in proper context, can be used to refer to the material and often wild world in the modern sense.

It is no coincidence that "nature (i)" in the English language, with its origin in the Latin word *natura* (birth, quality), is used to denote the inherent character of something, something intrinsic rather than external, something "of its own accord" rather than depending on external forces. Here, we see the linkage between "nature (i)" and the Daoist *ziran*. When we say "it is human nature to do something" and when Laozi says "it is *ziran* to say few words" (Lin, *The Wisdom of Laotse* 140), nature and *ziran* are used in the same sense, indicating something innate, built-in and independent, rather than something designed or contrived by human intention. In the Daoist philosophy, it is human nature to live a *ziran* life, following the heaven and the earth and ultimately manifesting *Dao* itself. Nor is it coincidence that *ziran* is used to refer to "nature (iii)" in modern Chinese. When Laozi sets the "modeling" sequence in *Dao De Jing*, Heaven/sky and Earth, and everything between them, are exactly what we call "nature (iii)" in modern languages, modern English, modern Chinese or any other modern language. Therefore, it is totally legitimate in the modern context to translate Laozi's *ziran* as nature (iii), even though he used the word with a much more metaphysical sense than mere nature (iii). Chen Shao-ming did not exclude "nature (iii)" in his introduction to *ziran*: "The word is only used with reference to all kinds of natural phenomena such as wind, rain, waters and seas, valleys, sounds, and colors. It suggests that one should follow the natural course or let nature take its own way." (97)[4] The "natural phenomena" are exactly what nature denotes in the modern sense, so the "naturalness" or "spontaneity" meaning of *ziran* and "nature (iii)" are in semantic harmony in the Daoist tradition.

This usage of *ziran* as "nature (iii)" is strengthened by the connotations the two words invoke in the philosophical sense. In the Daoist discourse, *ziran* is clearly used to denote things natural, as against human cleverness, avarice, and pretentiousness. In the anthropocentric mindset and practice, humans have long diverged from the natural way of life they once shared with other animals in nonhuman nature. They kill not just to eat; they produce to waste; they consume too much. The human way of life has threatened the survival of other beings that they share the planet with; they have become a geological force—that is the very definition of the Anthropocene. Robert Fletcher points out that our conception of human–nature relationship is "a deep-seated ambivalence wherein humans are thus described as simultaneously part of yet separate from this thing called 'nature' they are supposed to reconnect with" (229).

While Laozi's usage of *ziran* can only be rendered as nature (iii) with careful analysis and notation, Zhuangzi was almost outspoken when he was depicting the Daoist utopian world:

> Therefore, in ancient times when perfect virtue prevailed, the people walked in self-contentment and did not care to look around. At that time, there were no trails or paths in the mountains and there were no boats or bridges on the waters. Everything thrived on earth, with no barriers between each other. Birds and animals prospered; grasses and trees grew luxuriously. Under these circumstances, the birds and animals could be tethered and led about. People could even climb up to the magpies' nests and peep into them without disturbing them. In ancient times when perfect virtue prevailed, people lived together with birds and animals and mixed with everything in the world. How did they know the distinction between superior men [gentlemen] and inferior men [commoners]? All ignorant, they did not lose their virtue; all desireless, they were in a state of natural simplicity as uncarved timber, which kept intact their inborn nature. (Zhuangzi 137)

Zhuangzi's description of this "ancient times when perfect virtue (*de*) prevailed" can be seen as nostalgia of good old days when people lived in a *ziran* manner. Classical Chinese literature had the tradition of imagining an ideal time and an ideal king in the past, from Confucianism's *datong* (a world of universal harmony) to Laozi's "a small country with a small population." Idealization of this kind is often seen as lost because the ruler or the people did not conform to the teaching of the specific school, but also achieved if the ruler and the people do. In the historical context, Laozi and Zhuangzi's time was already long into the Anthropocene, if the Agricultural Revolution is taken as its starting point, so Laozi's "a small country with a small population" and Zhuangzi's "ancient times when perfect virtue (*de*) prevailed" can be both seen as their different reaction to the already apparent symptom of "the minor Anthropocene."

In Zhuangzi's mind, the cultivation of *de* is in accordance with *Dao* and *ziran*. A nature writer or a modern environmentalist would be thrilled to see such a life where humans and other beings share nature like they once did. The mountains and waters keep their pristine state; the habitats of wildlife are not yet fragmented; anthropogenic extinction is not even close; and most important of all, humans and other creatures live harmoniously together. This Daoist utopia is obviously a depiction of the bygone Golden Era that might or might not have existed in real history, but some aspects of it, especially those about the abundance of wildlife, the vastness of their habitats and the simplicity of humans, are of historical relevance. All humans, like any other beings in the same shared nature, are equal to each other without a Confucianist hierarchy of gentlemen and commoners.

In the "modeling" sequence Laozi sets for the "Four Great"—*Dao,* Heaven, Earth, and humans (Lin, *The Wisdom of Laotse* 145), and also in Zhuangzi's utopian imagination of "ancient times when perfect virtue (*de*) prevailed," humans clearly exist as part of nature, as a modern ecocritic would probably envision for an ecological utopia. But when coming back to reality, Laozi made the distinction between the Way of Heaven and man's way: "It is the Way of Heaven to take away from those that have too much and give to those that have not enough. Not so with man's way: He takes away from those that have not and gives it as tribute to those that have too much." (Lin, *The Wisdom of Laotse* 306) Nature usually maintains a dynamic equilibrium between its composing elements, plants, animals, and other lives share the same lithosphere and atmosphere. But man's way is to take more, pushing many other species to extinction and driving the climate to the brim of irreversible warming. Rich countries and rich class in poorer countries, primarily to blame for the changing climate, are the least affected in the Anthropocene, while underdeveloped countries and disadvantaged groups that have created fewer problems are those who will suffer the most in the coming catastrophe. Both the cross-species and inside-human-species injustices directly result from man's way that has gone astray from that of nature.

In *Zhuangzi*, likewise, when asked by the River God to differentiate between the (Way of) Heaven and (the Way) of humans, the Sea God replied, "That the oxen and the horses have four feet is what I mean by their inborn nature [the Way of Heaven]; that the horses are bridled and the oxen are led by the nose is what I mean by their enforced [*sic*] behavior [the Way of Humans]" (Zhuangzi 273). Bridling of horses and nose-ringing of cattle convict humans of an ecological crime that violates the innate nature of horse and cattle as independents species, but humans are happy to commit this kind of cross-species crimes, obviously to serve their economic gains. Physical harm and psychological torment all happen in the name of development and progress, in a process of slow violence, numbing people's senses and critical thinking. It is "a violence that occurs gradually and out of sight, a violence of delayed destruction that is dispersed across time and space, an attritional violence that is typically not viewed as violence at all" (Nixon 2). It is also a "structural violence" that is deeply rooted in the social, economic, political and cultural superstructure of the Capitalocene. The Anthropocene, with its associated calamities, is not some ailment that can be relieved by minor changes like full-belly environmentalism. It requires fundamental changes in the deep structure of human culture that will revolutionize the human way of life. Put in the modern context, Laozi and Zhuangzi's message is for us to rethink human aspirations, social ambitions, and ultimately "man's way."

The call for returning to nature's way, or the Way of Heaven, is most evident in Zhuangzi's preference to be a turtle that "drags its tail in the mud" (Zhuangzi 281) rather than a sacred turtle that serves as noble sacrifice in the state of Chu. In the Daoist context, the two options also represent the choice between a natural life and an anthropocentric life we as humans have to face. The anthropogenic climate change that will soon be irreversible in eleven years according to the latest IPCC report attests to the forecast that the anthropocentric lifestyle is unsustainable and will soon reach a *cul-de-sac*. "Some impacts may be long-lasting or irreversible, such as the loss of some ecosystems (*high confidence*)" (IPCC 7). "The risk of irreversible loss of many marine and coastal ecosystems increases with global warming, especially at 2°C or more (*high confidence*)" (IPCC 10). Zhuangzi's choice points to an alternative we can follow for the benefits of our own species and for billions of other creatures in nonhuman nature.

THE *ZIRAN* THERAPY FOR THE ANTHROPOCENE

From a philosophical perspective, the Daoist *ziran* works against the anthropocentric objectification of nature and reasserts the subjectivity of nature. The notion of nature's subjectivity has deeply influenced the Daoist approach to the non-human world. "Rather than understanding human beings as 'subject' who observe the 'objective' world of nature, Daoism proposes that subjectivity is grounded in *Dao* itself, the wellspring of cosmic creativity for a world of constant transformation" (*China's Green Religion* 26). This differs far from the modern humanistic imagination of the individual self and subjective agency that have resided in the cognitive capabilities of humans. The humans' (almost) unique abilities to think freely and creatively separate them from nonhuman nature, help them win the natural competition among all species, and in our hubris, make us the more advanced species. But in the Daoist cosmos, the capability of humans to think and act self-consciously does not separate them from nature; rather, it is nature that has endowed humans with such capability. A quick recap of the Daoist creation and modeling sequence would reveal this. *Dao* gives birth to everything, obviously including humans: "Out of Tao, One is born; Out of One, Two; Out of Two, Three; Out of Three, the created universe" (Lin, *The Wisdom of Laotse* 214). *Dao* and the natural world it creates are creative process of self-emergence and subjective agents that embody a mysterious virtue (*de*)—that is, *ziran*. Therefore, humans by birth would follow Heaven and Earth, which represent nature in general.

The Western idea of an inert and objective nature gives rise to technological imperialism and scientism, in which humans hubristically bring all nonhuman nature under control for their own benefits. The Daoist view offers an alternative: *Dao* ultimately gives ground for subjectivity. The human possession of subjectivity, a sense of self, an identity, and self-consciousness comes precisely from the fact that they emerge from nature and model after natural Heaven and Earth. "That humans possess autonomy and subjectivity is not something that makes them different from nature but rather is precisely what they inherit from the earth" (Miller, *China's Green Religion* 32). In the modern scientific context, this nature subjectivity actually makes a lot more sense than the Western human subjectivity, which suits the Judeo-Christian Genesis framework. James Miller proposes to look at nature from the perspective of evolution. "When understood as a process of emergence over 13.7 billion years, the raw power of nature from the big bang to the evolution of life on earth can be seen as nothing but sheer, overwhelming subjectivity" (*China's Green Religion* 33). From a truly scientific point of view, it is hard, even impossible, to separate humans from nature simply because humans have self-consciousness, feelings, and free thinking. Humans as bones and flesh, and their very possession of self-consciousness, are products of an ongoing natural process. The idea of nature subjectivity explains the Daoist reverence for *ziran*. The *ziran* subjectivity defines *Dao* itself and is an endowment *Dao* instills in every creature. Therefore, "the Daoist goal is to avoid doing anything that might inhibit the subjective agency inherent within things" (*China's Green Religion* 33).

From a practical point of view, *ziran* sets an ecological ideal for humans to live in real life. In history it has infiltrated into every aspect of the Chinese life and the Chinese mindset. "Although these early Daoists may not have been environmentalists in the modern sense, their religious worldview was founded on a cosmic ecology whose ideal state was a dynamic homeostatic equilibrium" (Miller, "Ecology and Religion" 2636). This equilibrium is first of all an equilibrium between humans and nature, but it is also an equilibrium between humans themselves and even an equilibrium within humans themselves. When humans stop advancing too much into nature and exploiting nature, when they mingle the human lifestyle with nature, the ecological crisis can be resolved or would not have arisen in the first place. When humans adopt the *ziran* way of life, the exploitative mindset will stop working not only for nature but also for the different people, social injustice will be neutralized. But ultimately, the *ziran* ideal works on the human mind. Only when humans truly regard the *ziran* equilibrium with nature as an ideal state of life, can behavioral changes really happen.

The Daoist ideal of *ziran*/nature has highly influenced Chinese poetry, art, and the Chinese way of life in general. Wei Qingqi has aptly summarized the embodiment of nature in all aspects of the Chinese lifestyle. "To the Chinese mind, nature is the embodiment of the *Dao* and therefore always simple, harmonious and pleasant, while social construction is most of the time unnatural, overcomplicated, and more or less awkward" (Wei 777). This ideal harmony between nature and humans is evidently shown in Chinese landscape paintings, which either foreground the mountains and waters or depict mingled scenes of natural beauty with human presence assimilated in nature. Robert Macfarlane regards Chinese landscape paintings as "an art in which almost no divide exists between nature and the human" (Macfarlane 2008:32). So is the case for classic Chinese poetry. Classical Chinese poets believe that "all of nature, whether it be mountains, rocks, streams, or trees, was[is] indeed alive" and "each element [has] an individuality that spoke of something more than just what it [is] on the surface each had[has] qualities which were[are] characteristic of the principles of life." (Lewis 7) They don't see nature as "a physical manifestation of its Creator, as it is to Wordsworth but something that is what it is by virtue of itself," so humans are "not conceived of as forever struggling against Nature but forming part of it" and are "advised to submerge his being in the infinite flux of things and to allow his own life and death to become part of the eternal cycle of birth, growth, decline, death, and re-birth that goes on in Nature." (Liu, James J. Y. 49)

The Daoist *ziran* has also permeated into the general way of life and way of thinking of the Chinese people. For example, the simplest principle of *fengshui* requires a house to be built near a mountain and a body of water. In Lin Yutang's words, China, along with its people, "has received her share of nature's bounty, has clung to her bowers and birds and hills and dales for her inspiration and moral support, which alone have kept her heart whole and pure and prevented the race from civic social degeneration" (Lin, *My Country and My People* 5). This defining feature of the Chinese lifestyle to try to stay close to nature also influenced their attitude toward femininity. Lin observed the fact that "women's dress is not designed to reveal the body of the human form but to simulate nature" (Lin, *My Country and My People* 143), and on a more philosophical level that "While in Western art the feminine body is taken as the source of inspiration and the highest perfection of pure rhythm, in Chinese art the feminine body itself borrows its beauty from the rhythms of nature" (Lin, *My Country and My People* 142). He therefore pointed out the correlation between the Chinese mind and the femininity: "Indeed, the Chinese mind is akin to the feminine mind in many respects. Femininity, in fact, is the only word that can summarize its various aspects. The qualities of the feminine intelligence and feminine logic are exactly the qualities of the Chinese mind" (Lin, *My Country and My People* 76). Similar discourse of feminine and natural as against masculine and scientific did not happen until the second wave of feminism in the 1970s. In his latest article, Wei also identifies that "the Chinese are prone to think of nature from a holistic and intuitive way rather than breaking it apart and dissecting its different parts" (779).

CONCLUSION

Written in "the minor Anthropocene" when human impacts on the planet have shown symptoms, which apparently have been grasped by Laozi and other Daoist masters, Daoist texts have an undertone similar in spirit to modern ecocriticism. Daoism, through its key concept of *ziran*, has proposed a set of cosmological order different from that of the postindustrial anthropocentric reality. *Ziran* and the whole Daoist "green" philosophical might very well serve as a therapy for the Anthropocene.

NOTES

1. The International Geosphere-Biosphere Programme (IGBP) was a research programme that ran from 1987 to 2015 dedicated to studying the phenomenon of global change.

2. *Wuwei* literally means "inaction" or "non-interference," but this concept in the Daoist concept would lean closer to "responsible action," not human-centered action but holistically ecological action.

3. Anthony Quinton's "Knowledge and Belief" in *Encyclopedia of Philosophy, Second Edition* analyzes Protagoras's famous thesis that "man is the measure of all things" from an epistemological point of view, but in ecology, this quote of Protagoras is more often interpreted as an anthropocentric claim that human needs dictate the value of non-human nature.

4. Chen Shao-ming's "Zi Ran (Nature): A Word that (Re)constructs Thought and Life" traces the origin of *ziran* and interprets this key term from an ecological perspective. I use Chen's interpretation to link the two interpretations of *ziran*, the original meaning of "self-so" and the modern translation of objective nature.

WORKS CITED

Chen, Guying. *Laozi, Annotations and Comments*. Beijing: Zhonghua Book Company, 1948.

Chen, Shao-ming. "Zi Ran (Nature): A Word that (Re)constructs Thought and Life." In *Keywords: Nature: For a Different Kind of Globalization*. Ed. Nadia Tazi. New York: Other Press, 2005.

Elvin, Mark. *The Retreat of the Elephants: An Environmental History of China*. New Haven and London: Yale University Press, 2004.

Fletcher, Robert. "Connection with Nature is an Oxymoron: A Political-Ecology of 'Nature- Deficit Disorder.'" *The Journal of Environmental Education* 48.4 (2017), 226–33.

Hardin, Garrett. "The Tragedy of the Commons." *Science*. New Series 162.3859 (Dec. 13, 1968), 1243–48.

IPCC. *Global Warming of 1.5°C. IPCC*. https://report.ipcc.ch/sr15/pdf/sr15_spm_final.pdf.

Lewis, Richard. *The Luminous Landscape: Chinese Art and Poetry*. New York: Doubleday, 1981.

Lin, Yutang. *My Country and My People*. London and Toronto: The Windmill Press, 1936.

Lin, Yutang. *The Wisdom of Laotse*. New York: Random House, Inc., 1948.

Linzey, Andrew and Clair Linzey. "Introduction: Toward a New(er) Religious Ethic for Animals." In *The Routledge Handbook of Religion and Animal Ethics*. Ed. Andrew Linzey, Clair Linzey. New York: Routledge, 2018, pp. i–xxv.

Liu, James J. Y. *The Art of Chinese Poetry*. Chicago: University of Chicago Press, 1966.

Liu, Xiaogan. "An Inquiry into the Core Value of Laozi's Philosophy." In *Religious and Philosophical Aspects of the Laozi*. Ed. Mark Csikszentmihalyi and Philip J. Ivanhoe. Albany, NY: State University of New York Press, 1999, pp. 211–37.

Macfarlane, Robert. *The Wild Places*. London: Penguin Books, 2008.

Malthus, Thomas. *An Essay on the Principle of Population*. London: Electronic Scholarly Publishing Project. http://www.esp.org, 1998.

Miller, James. "Ecology and Religion: Ecology and Daoism." In *Encyclopedia of Religion, Second Edition*. Ed. Lindsay Jones. New York: Thomson Gale, 2005, pp. 2635–38.

Miller, James. *China's Green Religion: Daoism and the Quest for a Sustainable Future*. New York: Columbia University Press, 2017.

Nixon, Rob. *Slow Violence and the Environmentalism of the Poor*. Cambridge, MA: Harvard University Press, 2011.

Paul J. Crutzen. "Geology of Mankind." *Nature* 415.6867 (2002), 23.

Quinton, Anthony. "Knowledge and Belief." In *Encyclopedia of Philosophy*, 2nd ed. Ed. Donald M. Borchert. Farmington Hills, MI: Thomson Gale, 2006, pp. 91–100.

Ruddiman, William. "The Anthropogenic Greenhouse Era Began Thousands of Years Ago." *Climatic Change* 61.3 (2003), 261–93.

Wei, Qingqi. "Toward a Holistic Ecofeminism: A Chinese Perspective." *Comparative Literature Studies* 55.4 (2018), 773–86.

White, Lynn, Jr. "The Historical Roots of Our Ecological Crisis." *Science* 155.3767 (1967), 1203–07.

Williams, Raymond. *Keywords—A Vocabulary of Culture and Society*. New York: Oxford University Press, 1985.

Zhu, Kezhen, "Zhongguo jin qiannian lai qihou bianqian di chubu yanjiu" (A preliminary study of climatic changes in China during the last 5000 years). *Kaogu xuebao*, 1 (1972), 15–38.

Zhuangzi. *Zhuangzi*. Trans. Wang Rongpei. Changsha: Hunan Peoples Publishing House, 1999.

Chapter 4

Toward an Ecocriticism of Cultural Diversity

Animism in the Novels of Guo Xuebo and Chi Zijian

Lili Song

There is no denying that the global environmental crisis brought on by pollution, climate change, the nuclear risk, and the depletion of energy and resources threatens the very future of humanity. These have resulted from the exploitation of natural resources facilitated by advanced science and technology, and driven by the insatiable appetites of modern industrialized urban societies. Predicament and threat therefore prompt us to reflect on cultural changes both locally and globally. Historically, the relationship between people and place in Chinese context has undergone a process of drastic change that has seen traditional values and beliefs, once closely related to the local environment, be ripped apart and swept away by modernization and globalization, both of which tend toward unified values and serve to encourage people's belief in a better material life. While nothing is inherently wrong with such a belief, the sustainability, health and spirituality of humanity, which depends on a symbiotic relationship with the natural world, cannot be simply disregarded. Addressing these contemporary environmental problems, which reveal a patently disruptive and unpromising human existence in the natural world, is thus to reflect upon the relationship between nature and humanity above and beyond modernization and globalization. Equally necessary is a reexamination of the environmental significance of traditional cultural value and belief systems, which have generally been constructed upon the condition of interaction and interrelationship between man and "local place."

The fact that it is the geological diversity of living conditions which creates the multifarious nature of culture and literature has been reemerging within the scope of ecocriticism, particularly in its third wave,[1] where the study of

71

place and environmental justice brings the issues of race and culture into consideration and encourages the study of minority or indigenous literature and culture. According to Sarah Pilgrim and Jules Pretty, "Even when considered as a dichotomy, it is clear that nature and culture converge on many levels that span belief system[s], social and institutional organizations, norms, stories, knowledge, behaviors and languages" (Pilgrim and Pretty 2010, 1). While biodiversity is now considered to be essential to the health and resilience of ecosystems, "cultur[al] diversity as a part of the interconnected whole" (Pilgrim and Pretty 2010, 1) responsively demonstrates its importance for humanity as we seek to cope with the environmental predicament. As T. V. Reed suggested in his article "Toward an Environmental Justice Ecocriticism," different traditions of nature writing deserve similar attention (Reed 2002, 149). Now, therefore, it is naturally important for ecocriticism to move toward an ecocriticism of culture diversity in order to better address the predicament of and threats to the future of humanity.

Roughly, an ecocriticism of cultural diversity is the study of the relationship between nature and literature from any culture. Patrick Murphy initiated this trend by pointing out the necessity of inclusiveness and "reconsidering the privileging of certain genres and the privileging of certain national literature[s] and certain ethnicities within those national literatures" (Murphy 2000, 58). In his reconsideration, Murphy called attention to the need to draw energy and resources broadly to refresh and enrich the study of nature-oriented literature and the environment. Scott Slovic, in his article titled "The Third Wave of Ecocriticism: North America Reflections on the current phase of the discipline," claims that the field now "transcends ethnic and national boundaries" and "explores all facets of human experience from an environmental viewpoint" (Slovic 2010, 6–7). He traces the development of ecocriticism toward diversified cultural vision by outlining the features of multiethnic literary study (Ecozon@, 2010). He urges scholars to "consider the importance of retaining ethnic identities but placing ethnically inflected experience in broader, comparative contexts" (Slovic 2010, 7). This corresponds with the construction of ecoculture, which "comprise[s] human cultures that have retained, or strive to regain, their connection with the local environment, and in doing so are improving their own resilience in light of the multitude of pressures they face, including global climate change" (Pilgrim and Pretty 2010, 11). This chapter, for its part, seeks to go beyond the scope of globalized modernization in order to study animism and exemplify a cultural diversity ecocriticism by analyzing two works of Chinese minority literature.

Any approach to the subject of cultural diversity will first of all involve a reexamination of the values and beliefs generated through long-term interaction with the local place, as these existed before they were suppressed by global modernization and condemned as backward and superstitious. China, with its diversified landscape and topography, is home to fifty-six ethnic groups whose cultures have been sustained despite marginalization and modernization. Minority literature in China therefore plays an essential role in an ecocriticism of cultural diversity that seeks to recollect local knowledge and reconnect humankind with the natural world. It has been a narrative tradition for Chinese minority literature to focus on the interaction and interrelationship between people and their environment based on the fundamental belief in animism that takes a distinctive form in each culture. For example, Ethnic *Miao* take animism as their ethical foundation, establishing a special relationship with the forest, and animistic beliefs are reflected in its architecture (He and Xia 2010, 22–25). The Miao housing style expresses the equality between humankind and everything in the natural environment. Likewise, many northern minorities such as *Manchu, Oroqen, Ewenki,* and *Mongolia* (before it turned to Tibetan Buddhism), and *Uygur* (before it turned to Muslim), took animism as the basis of their belief as expressed through totems, nature, and ancestor worship to show their respect for natural beings and to maintain the balance between humankind and the local ecology (Chen 2007, 150). Ethnic *Yi* and *Bai* also trace their tribal origins to certain mysterious wild animals, worshiping them as their totem and believing that there exists a supernatural connection that can never be completely erased (Lei 2009, 127–128). Even Tibetan Buddhism contains a certain idea of animism that can be inferred when reading the classic epic *King Gelsall,* in which the evil power cannot be conquered unless its controlling spiritual animal is killed. So animism is an essential issue to be addressed in the study of contemporary environmentally oriented minority literature.

Animism could be a touch point for reviving the vitality of cultural diversity. We have to acknowledge that, in the modernistic world view, animism is dismissed as irrationally other, and seen as a marker of primitivism. The modern objectivist world view, particularly when supported by the development of science and rationality, has no room to tolerate an inspirited connection with the natural world. Anthropocentric modernity has been cutting down and eliminating animism as the root of superstition and backwardness in the hope of making the subjectivity of humanity scientific, but at the same time it is also isolating humanity from its connection with the natural world. However, prior to the advent of globalized modernization, animism had informed the worldviews of many indigenous and regional ethnic groups and played a fundamental role in shaping their culture and literary representation, in which there was no dividing line between nature and culture. It has been deeply embedded in culture and literary representation, neither has it ever been totally erased from their lives. Anselm Franke stated in the introduction to *Animism* that animism is still alive (approximately 40 percent of the world's population believe in animism), and that it represents the fundamental "antitype" to the western world view (Franke 2013, 3).

Animism means that all natural beings, even the whole universe itself, are considered to be alive and inspirited. Chinese ethnologists have demonstrated that animism is a belief in a world with spirits, that interact with the human world, and now and then exert good or bad influence (Guo 2001, 69). Animistic cultures believe that there is a connection between the inspirited natural world and the materialized natural world, and their myths and tales perceive and represent this connection. The perception and representation of this connection is the characteristic element of the relationship between culture and nature. Through the practice of shamanism, the idea that the world is an interconnected and transformative living sphere gradually leads to the creation of the shamanic cultural rituals of nature worship, centering on sky, mountain, land, forest, river, etc. These display a distinctive understanding about nature and about how human beings can gain the blessings of a healthy and safe life. As Christopher Manes states, "to regard nature as alive and articulate has consequences in the realm of social practices. It conditions what passes for knowledge about nature and how institutions put that knowledge to use" (Manes 1996, 16). Animism acts as the elementary belief system for many ethnic groups all over the world, including those practicing nature worship and ancestor worship. Likewise, ethnologist Xingliang has shown how natural worship based on animism includes the worship of sky, earth, sun, moon, stars, thunder cloud, rainbow, wind, cloud, mountain, water, fire, etc. Each form of worship creates diversified cultural elements, among them ideology, name, imagery, worshiping place, ritual, taboo, and legends (He 2008, 3). He suggests that the diversity of culture must be understood in the context of analyzing not only spatial distance but also chronological evolution, including its origin, development, its inclusiveness, and its transformation (He 3). This, in turn, would help us to think about where to pick up our broken relationship with the natural world.

Diversified worship practices lead to diversified cultural creation and construction, which contains an alternative knowledge of the interaction between humankind and the natural world. This chapter seeks to demonstrate the value of an ecocriticism of cultural diversity by exploring the implications of animism in shamanic culture as represented in two Chinese minority novels that examine both environmental problems and alternative conceptions of the human relationship with the world. I will focus on Guo Xuobo's *The Golden Sheep Cart of the Leading Shaman* (abbreviated as GSCLS in the following text), which tells the story of a search for the Mongolian shaman against the backdrop of environmental deterioration, and Chi Zijian's *The Right Bank of Argun* (abbreviated as RBA in the following text), an account of how the Ewenki people interact with the forests, bears, and deer in their traditional way of life. The following analysis will begin with Guo's narration of the search for the shaman and its culture before going on to examine its underlying worldview of animism.

NARRATING THE SEARCH FOR THE
SHAMAN AND SHAMANIC CULTURE

The search for shamanic culture has been Guo Xuebo's distinctive narrative perspective in his shaman story series, in which the narrator is always the "I" looking for the ancient shaman culture long thought to have disappeared from northern China in the wake of modernization. Each story develops against the background of a local environmental problem, such as mine drilling, desertification, or the loss of prairie. This narrative perspective is closely related to Guo's Mongolian ethnic background, which, like other nomadic groups in northern China such as Manchu, Ewenki, *Hezhe*, Uygur, and *Kazakh*, has a long shamanic tradition rooted in animism. On account of the belief that everything is spirited, and the rituals of nature worship, totem worship, and ancestor worship, there are imperatives within these ethnic groups that humankind must respect nature, venerate all forms of life, and observe a variety of taboos, which traditionally serve to adjust the relationship between local peoples and their environment. Shamanic culture therefore developed on the basis of a belief in animism that carries with it an eco-ethical value enabling people to adapt to the living environment and maintain a symbiotic relationship with their surroundings (Chen 2007, 150).

In recent years, with the deterioration of the environment and the questioning and criticism of modernity, minority literature has emerged as a strong force committed to exposing the reality of ecological crisis and to exploring its origins. Authors are returning to their own ethnic roots, and recollecting their traditional beliefs and values (Lei 2009, 126), including shamanism and animism, which coincide with the flourishing of environmental literature (Yuan 2014, 68). Chi Zijian and Guo Xuebo are the most distinguished exponents of this trend. Guo, one of the few Chinese authors who claim to be an authentic environmental writer, stated in an interview. "Since I grew up on the desertified prairie and witnessed the progressively worsening process of desertification, my writing naturally touches upon and confronts the problem of environmental reality" (Yang 2010). In the ten novels he has published, among them *Silver Fox*, *Sand Fox*, and *Wolf Child*, the connection between the search for shaman culture and concerns about environmental degradation have always been the main intertexture of his literary representation of humanity and the environment. Confronted by environmental exploitation and deterioration such as desertification of the prairie and mountain mining, the stories' narrators embark on a journey in search of shaman culture. The stories usually end with the narrator finding the shaman engaged in the practice of worshiping rituals, discovering the shaman's knowledge of the cultivation of new crops, or his teaching of the shaman songs. In metaphorical terms, all of these convey a message of the writer's effort to reconnect with shaman culture in order to search for a way to achieve a sustainable life. This process of reconnection prompts readers to reflect upon the contemporary potential for the revival of the shaman cultures that have been either marginalized or destroyed. Guo's novel contains six different and independent shaman stories linked by the narrator and by a distinct message of the need to protect nature. Taking the first story in the book as example, the shaman, whom the narrator finally manages to track down, is found leading a ritual designed to protect the mountain from being mined by the combined forces of local government and greedy miners. According to the author, these six shaman stories are his most precious creation, from which we infer the literary message that shaman culture is to be revived while environmental degradation progresses, so that we may create a different view of the relationship between humankind and nature.

THE ANIMISTIC WORLDVIEW OF SHAMAN CULTURE

The first story in GSCLS commences with a shaman prayer song entreating Blue Sky, Tenger to "let the happiness, fortune flow from heavenly storage to the human world, and to grant them to the common people, and let the cow's and sheep's milk flow as spring water, and let grains fill up the land and let happiness permeate mankind, oh merciful Sky Father" (Guo 1–2). On the surface, this song is no different from other utilitarian religious prayers petitioning the deity for blessing or good fortune. But in this prayer song, the Mongolian-imagined world is centered not on the God, a metaphysical father, but on the Blue Sky, Tenger, a spirited physical father. The sky is a common center for animism worship in shamanic cultures, yet, everything that forms a part of nature—the sky, land, mountain, river, and natural phenomena such as wind, rain, thunder, and lightning—has its own spirit and is closely associated with human behavior. This may sound epistemologically primitive when compared to the modern scientific worldview, but it resulted in a sustainable relationship with the natural world by means of which human complacency was held in check and human damage to the diversified natural existence constrained.

The animism worldview, which originated from ancestors' direct observation and visualized imagination of the world, regards the world as both a spiritually oriented ontology and a physically interconnected whole, which means that it is very unlikely to be understood from the oppositional subject-and-object view. Humankind cannot detach itself from a spiritual world centered around the sky (or other natural object) and has been strongly affected in one way or other by the interaction between the spiritual world and human activity in the physical world (Guo 52). Reverence for and worship of the natural objects thus appear to be the essential core of animism belief, which in turn leads to the establishment of shaman culture. The shaman acts as the negotiator of the consequences of the interaction, for example, healing diseases, saving people from natural disasters, or simply assuring the material satisfaction of daily life. In this prayer song, we see that the shaman sings as if he were communicating with Tenger, the physical father, interceding on behalf of the welfare of the common people. The section of this chapter that follows next involves a closer reading of the shaman story in the literary texts of both Guo and Chi by focusing on the three characteristics of the relationship with the animistic world: communicative, transformative, and holistic.

COMMUNICATIVE

The communicative characteristic of the animistic relation with the world is illustrated predominantly through shamanic ritual. In GSCLS, the *Wengedu* Mountain Stone Material Factory celebrates its opening by blowing up the *Wengedu* Mountain. In Inner Mongolia, as well as in northwestern China, numerous stone material factories sought to make quick money from the urban construction and real estate development boom, causing great destruction to the natural scenery and environment. Developers of the factories often obtained permission to mine by bribing government officials rather than by complying with the administrative procedures and cultural considerations governing production and consumption. This, of course, is a practice that puts profit before everything else. While profit making plays the ultimate role in the process of Chinese modernization and industrialization, the centralized government, which is rife with corruption, abuses its power to turn the whole of society into a mercenary machine while also implicating science and technology in the process. The natural environment has been targeted and victimized to pursue mercenary and consumptive development in a heedless and relentless manner. In Guo's description, the Wengedu Mountain was historically a place for shaman worshiping. The Obo, a heap of round stones at the top of the mountain, was erected there in the legendary story of the "Golden Sheep Cart." In local shaman culture, the Wengedu Mountain has been respected as an original holy place, a place with a soul and spirit, in other words, imbued with life rather than simply being an inorganic natural object. Although orders to have the Obo removed were issued several times when

Lamaism replaced shamanic belief, local history records show that the round base stone of the Obo could never be completely eradicated. In more recent years, Obo worshiping has been reviving. Instead of going to the temple, local people come to the mountain every fall with huge stones, and the Obo has been rebuilt and is becoming bigger and bigger. Now driven by economic gain, Nuke, the company executive and developer, bribed the local officials with money and sexual favors to obtain permission to open the factory by dynamiting the mountain for stone, the basic construction material required in the rapid process of urbanization and real estate development—with complete and utter disregard for the cultural significance of the mountain. The aged shaman, supported by the local villagers, decides to perform a ritual to protest against the destructive mining of this holy and culturally sacred mountain. In the ritual, the ninety year old shaman dressed in shamanic costume, wielding sword and cow bone bell, steps out of the golden sheep cart. He sings:

> the Bo (1) being descended from ancient ancestors,
> has been worshipping the spirits, from generation to generation,
> of Sky, Earth, Sun, Moon, Stars, and Every Natural Being!
> The majestic and auspicious Wengedu mountain,
> is a holy place for Tenger·Hailaerji—
> Where we can receive eternal love from the eternal sky. (Guo 55)

As the ritual is celebrated, the author presents the shaman as if he were communicating with the world of spirits and serving as a medium of connection between the local people and the mountain or, rather, the natural beings, which, in the shaman's ritual, are as life-infused as humans. Different spirits are depicted through the shaman's varying poses and gestures. The presentation of the spirits of the natural world shocks people into a sense of awe and respect, through which they are drawn into the celebration, becoming a part of the communication and starting to hymn and dance with the shaman. In the communal dance, the anger of the natural world is directed toward the destructive exploitation of the mountain, evidently expressing the wish of the local people for a safe and just life. At the same time, the spirits of nature and humankind seem to coexist, mutually responding to each other's demands.

The communicative connection has an imperative ethical implication of "measure for measure" in the sense that impetuous and heedless action on the part of humans may inflict punishment, serving as a demonstration of the link between humankind and the spirits of the natural beings. Following the shamanic ritual, the mountain is dynamited anyway and Nuke even disrespectfully plays with the shaman's golden sheep cart, shouting that he doesn't believe in so-called animism and proclaiming that he would himself step onto and run the dignified cart about to see what would then transpire (Guo 72).

However, while he is careering about wildly in the cart, an accident occurs. In the aftermath of the explosion, the huge Obo collapses and the black stone rolls down the mountainside, crushing the cart. Nuke dies together with the golden sheep cart. From the scientific point of view, it is simply an accident. But, looking at it from the interconnected perspective, can we simply treat the mountain as a source of construction material? Is the mountain as isolated from us as it appears to be? Is the human-and-mountain connection a mere subject-and-object relationship? In this shaman story, the author is suggesting negative answers to these questions. The whole story is indicating a worldview with communicative characteristics, as demonstrated through the punishment, which can be seen as the ecological moral implication of animism with an interconnected sense. According to Barry Commoner, the first principle of the ecological world is that "everything is connected with everything else." The destruction of the connection can lead to turmoil and holistic disaster. According to Taoism, the more intrusion into the natural world there is, the more trouble and suffering will be inflicted. Although animism expresses the connection in a deified and anthropomorphized manner, its eco-ethical implication realistically corresponds with the principles and wisdom of scientific ecology. Similarly, the connection between human action and the consequences of the changes made to the world is embedded in Mongolian and other Chinese minority cultures. Its moral ecological decrees are closely associated with environmental protection and serve as a warning, as expressed by the author: "We don't know how far we humans can go since we have lost our respect for nature, the sky and the earth."(Guo 74) This also echoes Rachel Carson's warning of a "Silent Spring" if human beings continue on their journey by further subjugating nature.

TRANSFORMATIVE

In shamanic practice, another striking characteristic that marks the human relationship with the world of animism is that life and spirit are transformative, particularly in the man-and-animal relationship. Anthropologist Seyin states that animism is such a compromising . . . way of thinking that it neither belongs to idealism nor materialism but the combination of both (Seyin 2011, 15). The world is an integrated world. The integrity lies in both the outer connection and the inner connection. In animistic belief, there is no oppositional separation but, rather, an interchange and transformative relationship among the lives of all beings in the natural world. Taking Chi's RBA as an example, we can read the spiritual connection between the life and death of human and natural beings throughout the story, most particularly between people and reindeer. This book narrates the story of a group of Ewenki nomads living with the reindeer in the forest before they were forced to settle down in the 1980s. Ewenki, like the Mongolian people, have a strong belief in animism and shamanism. In Chi's representation of Ewenki life, we see an inseparable world in which the forest, reindeer, bear, man, and woman are all connected, both materially and spiritually. The transformation and interchanges between the life and death of one another are understood as the connection between humankind and the world. Liena, the sister of the narrator, gets sick one autumn. Shaman Nidu is asked to perform a ritual to save her life. Nidu puts on his shaman costume and dances and sings in search of Liena's spirit. He dances and dances from dusk until the stars came out and, suddenly, he falls down in a faint. Liena wakes up and shaman Nidu tells their mother that a little reindeer's life has been transformed for Liena's life. They go outside and find that one of the little reindeer has died in the yard. The transformation of a reindeer's life for that of Liena indicates that, in Ewenki belief, all spirits are connected and human life is sometimes in debt to the life of other beings, and that other natural beings such as reindeer are therefore considered an inseparable part of human life. In Chi's *RBA*, we can see that, among the Ewenki, the reindeer were always taken as their family members, the bears were regarded as reverent elders, and the trees were considered as holy protectors.

The transformative characteristic of animism puts people and the living environment in an interconnected world, irrespective of material forms. The interconnection that comes with this belief implies a natural respect and equality, which keep Ewenki people from overexploitation of and an attitude of superiority toward the natural environment. Later in the story, when the tribe moves to another place to settle down, Liena gets sick again, and this time she drops off the back of the mother reindeer whose child once died for her. This episode can be interpreted as the mutual and equal relationship between a reindeer's life and that of an Ewenki. The mother reindeer is as emotional and cognitive as a human mother. It conveys the message that it remembers the pain of losing a child and claims the debt when the time comes.

In addition to the belief in equality among all beings, there is also a built-in sense of care that transforms from one life to another in the living Ewenki community. The narrator owes her first pregnancy to the water dog. She describes how one spring, she and her husband went to the river to hunt the water dog. They found four baby water dogs in the cave. Reflecting on the cruelty of taking them away from their mother, they refrained from killing the baby water dogs and left them alone. Their sympathy was transformed into a life reward and, a short time later, the narrator became pregnant after being married for three years. The narrator attributed the coming of the child to the caring attitude shown toward the water dog and therefore decided to stop hunting water dog. In Chi's rendering of the Ewenki story, transformative characteristics of the animistic worldview indicate that human and nonhuman life share a mutual bonding of life and spirit. In the mutual care between the water dog and the narrator, the separation of species is transcended, yielding a reciprocal benefit of protection. Christopher Manes states that: "Significantly animistic societies have, almost without exception, avoided the kind of environmental destruction that makes environmental ethics an explicitly social theme with us" (Manes 1996, 18).

Belief in the transformation of different spirits presupposes recognition of the eternity of the animistic world, in which individual human life is considered transient and unimportant when compared with the perpetuity of holistic existence. In Chi's novel, it is impressed upon us that time transcends human existence. The hunting and fishing activity marks the vitality of human life along the river, and at the same time fuses human existence with the movement of the natural world. Accordingly, the quality of life is measured not by the material possessions gained by manipulating and manufacturing but by the harmonious unity with natural beings gained through the lively interaction and interconnection with the environment, the river, the mountain, the forest, and the animals. Guo Xuebo also expresses this idea of eternity in animistic belief by observing that "the river flows to the east, eagles fly over Wengedu Mountain, everything we see changes and transfers but nature remains perpetual" (Guo 19). In Ewenki history, the people have maintained their way of life in the northeastern part of China and their belief in animism since 2000 BCE. Not until the middle of the twentieth century were they forced to leave the forest, the mountain, and the river to settle down. In little more than half a century, they have been faced with the total destruction of their culture and beliefs, while humankind as a whole is now confronted with the environmental and existential crisis resulting from factors such as desertification and global climate change. There inevitably appears to be a challenge to the modern humanistic standards applied when judging the advantages and disadvantages of culture, a challenge to the humanistic standards used to differentiate rational from irrational culture. There is also a need to question culture and humanity, to ask whether development should come at the cost of destroying the natural world upon which humanity depends, or whether development should be pursued with the restraint required to maintain the integrity of the natural world. The question could be put as follows: Should we sustain a caring and natural way of life, or engage in heedless, impetuous development? Chi Zijian wrote in the postscript that, long before modern people came to change the land and set about developing and constructing, Ewenki generated and regenerated on the northeastern land. "There is nothing wrong in culture developing. Does not God drop human beings onto the earth to search for the answer to existence? But the problem is that God asks us to search for a harmonious way of life rather than an exploitative and destructive existence" (Chi 2005, 252). When the sandstorms, desertification, and climate change came to us like a ghost, the fact of overexploitation and destruction questions the very direction modern cultural development and humanity is taking, and at the same time activates the memory of cultures long preceding the modern period that enable people to sustain their lives on earth.

HOLISTIC

In the animistic world view, humans see themselves as a subordinate part of the natural world, and the aesthetic ideal of human life is to make an effort to enter holistic existence. Humankind and everything in the natural world, either in material or in spiritual form, are inseparable, and are influenced by each other in one way or another. In other words, primitive holistic belief puts humans in a modest position to imagine and imitate the spiritual connection with the world. According to Zhang, animistic belief lays the foundation for premodern peoples to form an interdependent relationship with the natural environment, to intuitively understand the changes occurring in the natural world, and to find ways of unifying themselves with it in the hope of sustaining their production and reproduction rather than conquering and breaking it in order to make it serve human life (Zhang 2013, 88–89).

In Guo's novel, the holistic dimension of animistic belief is expressed by the depiction of ancient shaman-style singing, which is the surviving cultural element descended from traditional shamanic chanting. Three of six stories in it recount how aged shamans continue to cultivate this art, which Guo associates with the environmental crisis by saying that the tone of present-day Mongolian singing is mostly sad, as if weeping and lamenting, since the feeling of the song corresponds with desertification and the loss of the prairie. There are, in fact, double layers of sadness in contemporary shaman-style singing. For one thing, few people can now understand, or care to understand, shamanic chanting. For another, in today's world it is barely possible for shaman singing to have a happy tune. The reason for the former is that "modern people have been alienated from nature and blinded by their greediness for material benefit, and have thus lost their pure and natural heart and spirit" (Guo 87). The reason for the latter is desertification and the loss of home. As Guo writes, in the most recent one hundred years, with the decline of the belief in unity with the sky, earth, and everything in the natural world, the green grass has also been disappearing. The scenery, in which once "the sheep could only be seen in the grass when the wind blew the grass down," is now severely desertified and the pasture land so degraded that many local people are no longer able to live on it (Guo 81). There is a reciprocal relationship between the loss of the pure and natural spirit and the loss of the natural environment. Shaman-style music is to some extent considered to be a reaction to the holistic spirit of the natural world. In the old song titled "The Wind of the Sky," Guo's description of the sadness is penetrating, heartbroken, and soul-broken. In Guo's story, we discover that both the wolf and the narrator were captivated by the sadness of the song. However, the narrator, as an educated modern individual, could take it only as a beautiful sound, whereas the wolves howled with the song to give it a unifying ending and ran away. In other words, the old shaman concludes the song together with the howl of the wolf. The song sang out the holistic spirit of the world, of which both humans and the wolf form a part.

According to Guo, the song is not simply a piece of human production of sound art; it is a part and also a combination of the holistic spirit of the world. The singing originated from the imitation of various sounds in nature, among them the sound of a waterfall, mountain, forest, and various animals. Yet it is not just an imitation: it is a vehicle for the Mongolian people's understanding of the animistic world, and an effort to join in, to become a part of the animistic world. The singing expresses an effort to create the harmony that is made by all natural sounds. "The music expresses the Mongolian people's intuitive realization of partaking in the profound existence of the universe, and expresses their animistic idealism and aesthetic feeling" (Guo 115). In the process of imitation, the forefathers of the Mongolian people came to understand and cherish the livelihood of the world, thus divining the wisdom of how to act in accordance with the natural world. Their ancestor teachings include precepts such as: in spring, hunting is forbidden; tree branches are not allowed to break; never spit or pee into the river and lake; don't climb to the top of the mountain. When the old shaman teaches Arena, a pure and naïve girl, to learn how to sing the shaman-style song, he teaches her not only how to vocalize, but also teaches her the rules of nature and the importance of appreciating the vitality of the natural world as a whole. As Guo depicts, only by respecting and understanding the symbiosis of natural beings is Arena finally able to sing the incredible shaman-style music by joining in the holistic animistic world and being united with everything spiritually.

> She feels the impacts and communication with the eagle, a communication between souls. Her eyes shine with bright light and she experiences the incredible and ineffable resonance and communication between her own heart and the heart of the eagle, and suddenly feels as if she were soaring in the sky! Her singing is finally united with that of the mother eagle, thus creating a holistic harmony. Then her song reverberates between the sky and earth. The mountains and rivers, and every part of nature, are moved and listen silently to the harmony (Guo 120).

This is indicative of the author's effort to revive the shamanic expression of the holistic animistic worldview, in which man is related to the world by joining in the unity of the spiritual vitality of other natural beings. The greatest creation of human art is not to imitate natural beings but to join in the rhythm of natural life. That human beings form part of the natural world is contingent upon understanding the togetherness of spirits, which also judges the consequences of human action. It is the symbiosis that brings humans and everything else in the world together, thus demonstrating the ultimate achievement of happiness in human life.

The above analysis of shamanic rituals of worship, shamanic healing practices, and the depiction of inherited shaman song in the two novels has considered the man-and-nature relationship from an animistic worldview through its characteristic of being communicative, transformative, and holistic, thereby unveiling the eco-ethical and eco-aesthetic implications of shaman culture. Just as Barry Commoner said: "Nature knows best, and there is no such thing as a free lunch." Shaman culture also warns people of the punishment that will be made out for their relentless exploitation of spirited natural beings. This position reinforces the message contained in Aldo Leopold's land ethical ideas of extending love to the land community, not from an oppositional subject-and-object perspective but out of the holistic epistemology. Above all, however, spiritual unity with the natural world manifests itself in shaman culture, in both artistic creation and in personal happiness rather than through the conquering and possession of the natural world. In other words, spiritual satisfaction and purification carry more weight than material satisfaction due to the spiritual connection to the visible and invisible world.

In the modern age, the philosophy of the "control of nature," has guided the development of culture. However, as environmental problems of ever-greater severity—toxic pollution, soil erosion, desertification, climate change—are threatening human existence, there is a need to reexamine our prevailing worldview. There is also a need to be inclusive of other worldviews that have brought forth highly diverse cultural practices. As has been observed, "biodiversity is now recognized as a prerequisite to ecosystem health and resilience, as well as an essential precondition to sustainable livelihoods, human health, and many other social objectives" (Pilgrim and Pretty 2010, 2). Cultural diversity bears the similar capacity to maintain the knowledge, outlook, and reconnection with the natural world, and can effectively function as part of the "concomitant effort to sustain the inextricable connection" (Pilgrim and Pretty 3). Animism, once a preponderant worldview in many parts of the world, has spawned and supported environmentally friendly cultures, and the fact that it has never vanished, even in the face of sweeping rational modernization, proves that the inseparable tie with nature goes beyond anthropocentric epistemology, thus serving to override prevailing complacency about the dominant role of modern culture. With environmental threats increasing, and with the dilemma and paradox of modern consumption-based culture fast approaching a dead end, it is time for humanity to embrace diversified culture values and a belief system which is capable of proving itself valid for both the past and the present—and of safeguarding a sustainable future.

NOTE

1. "The third wave of ecocriticism" is first put forward by Scott Slovic in 2010 in his article "The Third wave of Ecocriticism: North American Reflections on the current phase of the discipline." In which he mapped the development of ecocriticism as three waves, the first wave mainly dealt with nature writing, wilderness, and modeled after feminism approach; the second wave dating from 1990s turned its attention to other literary genres and media, environmental justice and urban toxic issues, the third wave emerging with the twenty-first century the north American ecocritics started to adopt transcultural approaches to address the tension between global and local environmental issues and thus open the door to welcome the involvement of resilient relationship between man and of culture from culture diversity.

WORKS CITED

Adamson, Joni, and Scott Slovic. 2009. "Guest Editors' Introduction: The Shoulders We Stand on: An Introduction to Ethnicity and Ecocriticism." *MELUS* 34.2: 5–24.

Chen, Xu, Zhongguo Beifang Minzu Saman Wenhua Suo. 2007. "Tixian de Shengtai Lunli Jiazhi Guannian (The Eco-ethical Value in the Shaman Culture in Northern China)." *Zongjiaoxue Yanjiu* (Religious Studies) 2: 150–155.

Chi, Zijian. 1992. *Eergunahe Youan* (The Right Bank of Argun). Beijing: Beijing October Art Press, 2005.

Commoner, Barry. *Making Peace with the Planet*. New York: Pantheon Books.

Franke, Anselm. ed. 2013. *Wanwu Youling(Animism)*. Beijing: Gold Wall Press.

Guo, Shuyun. 2001 *Yuanshi Huotai Wenhua: Samanjiao Toushi* (The Primitive Living Culture: Shaman Perspective). Shanghai: Shanghai People's Press.

Guo, Xuobo. 2011. *Dasaman zhi Jinyangche* (The Golden Sheep Cart of the Leading Shaman). Beijing: New Star Press.

He, Minzhang, and Daiyun Xia. 2010. "Cong Wanwu Youling Kan Miaozu Jianzhu Zhongde Huanjing Lunli Sixiang (The Environmental Ethical Thinking in Miao's Architecture from Animism Perspective)." *Jishou Daxue Xuebao* (Shehuikexueban) (Journal of Jishou University (Social Sciences Edition). May 2010: 22–26.

He, Xingliang. 2008. *Zhongguo Ziran Chongbai* (Chinese Natural Worship). Nanjing: Jiangsu People's Press.

Lei, Ming. 2009. "Weiji Xungen: Minzu Wenhua de Rentong yu Xiandaixing Fansi (Looking for The Root while in Crisis: The Identification of Ethnic Culture and Reflection of Modernity)." *Qianyan* (Frontedge) 9: 126–130.

Manes, Christopher. 1996. "Nature and Silence." Glotfelty, Cheryll and Harold Fromm. eds. *Ecocriticism Reader: Landmarks in Literary Ecology.* Athens and London: The University of Georgia Press. 15–29. Print.

Meng, Huiying. 2004. *Xunzhao Shenmi de Saman Shijie* (Looking for the Mysterious Shaman World). Beijing: Xiyuan Press.

Murphy, Patrick D. 2000. *Farther Afield in the Study of Nature-Oriented Literature.* Charlottesville and London: University of Virginia Press. Print.

Pilgrim, Sarah, and Jules Pretty. 2010 "Nature and Culture: An Introduction." Pilgrim, Sarah, and Jules Pretty, eds. *Nature and Culture: Rebuilding Lost Connections.* London and Washington, DC: Earthscan. 1–20. Print.

Reed, T. V. 2002. "Toward an Environmental Justice Ecocriticism." Adamson, Joni, Mei Mei Evans, and Rachel Stein. eds. *The Environmental Justice Reader: Politics, Poetics, and Pedagogy.* Tucson: The University of Arizona Press. 145–162. Print.

Seyin. 2011. *Zhongguo Saman Wenhua Yanjiu* (The Study of Chinese Shaman Culture). Beijing: Ethnic Press.

Slovic, Scott. 2010. "The Third Wave of Ecocriticism, North American Reflections on the Current Phase of the Discipline." *European Journal of Literature, culture and Environment* 1.1: 4–10.

Yang, Yumei. 2010. "Shengming Yishi Yu Wenhua Qinghuai: Guo Xuebo Fangtan (The Life Consciousness and Culture Passion)." *Literary Newspaper*, July 5th issue.

Yuan, Jie. 2014. *Chuantong Samanjiao de Fuxing* (The Revival of Traditional Shaman Religion). Beijing: Social Science Academic Press.

Zhang, Shuping. 2013. *Shengtai Wenming Shiyu Zhongde Minjian Xinyang: Zhexinan Chuantong Xinyang Xisu Kaocha* (Local Belief In the Vision of Ecological Civilization: Examination of Traditional Custom and Belief in the Southwest of Zhejiang Province). Guangzhou: Jinan University Press.

Chapter 5

Population, Food, and Terraforming

Ethics in He Xi's *Alien Zone* and *Six: Realms of Existence*

Hua Li

In 1968, Paul Ehrlich used the term *population bomb* to refer to the explosive growth of the human population and its devastating impact on the environment. From his perspective, too many people with too little food to go around would result in a dying planet. However, "Plans for increasing food production invariably involve large-scale efforts at environmental modification" (Ehrlich 1968, 48). Overpopulation will lead to not only the deterioration of the environment, but also mass famines (15–67). In his next book *Population, Resources, Environment*, he further argues that "the most pressing factor now limiting the capacity of the Earth to support Homo sapiens is the supply of food" (Ehrlich 1970, 65). Ehrlich examines various strategies that human beings have deployed to solve the problem of food shortages, such as improving agricultural technologies, placing more land in cultivation, government-enforced population control policies, and mass migration to less populated regions. Though Ehrlich rejects the extrapolative idea of shipping the surplus population to outer space because of the limits of our current level of science and technology, various science fiction writers have delved deeply into the possibility of eventually transforming a lifeless planet into a habitable destination for human migration away from Earth. This type of fiction is typically referred to as terraforming fiction. Terraforming "involves processes aimed at adapting the environmental parameters of alien planets for habitation by Earth-bound life, and it includes methods for modifying a planet's climate, atmosphere, topology, and ecology" (Pak 1). Terraforming is not only a means for human beings to explore and expand their territories, but also becomes a site for environmental philosophical reflection. Terraforming "allows us to imaginatively re-situate our values with respect to our place in the universe, thus, calling for a re-evaluation of the assumptions behind vary-ing positions to nature and to each other" (Pak 7).

This chapter focuses on Chinese science fiction writer He Xi's (何夕 b. 1971) two terraforming narratives *Alien Zone* (*Yi yu* 异域, 1999) and *Six Realms of Existence* (*Liudao zhongsheng* 六道众生, 2002).[1] In the postscript of *Six Realms*, He Xi points out that the *Six Realms* is sequel to *Alien Zone*. Both works describe the consequences of human beings' excessive exploita-tion of nature. He also indicates that the writing of his postscript is to com-memorate the day when the global population surpassed the six billion mark (He 184). *Alien Zone* and *Six Realms* explore how humans might deal with shortages of both food and housing that have been triggered by increases in population. This type of narrative first appeared in China in the mid-1950s, became more widespread around 1980, and has especially surged in popular-ity since the dawn of the twenty-first century. However, He Xi's narratives about shortages in food and housing mark a departure from assumptions about there being an unlimited scope for the expansion of housing and agri-cultural productivity—an idea widespread in Chinese sf narratives from the mid-1950s to the 1980s. His novellas also mark a shift to self-restraint in development or smart growth. The two narratives also reject the longstanding approach to nature as ripe for plundering, and instead view nature as in need of healing; instead of viewing nature as being unrelated to ethical concerns, these narratives imply that humankind is indebted to nature and must behave more ethically toward it. In addition, the hypothetical terraforming projects in the two texts provide a venue to examine humankind's relationship with the vast and complex realm of nonliving space, as well as humankind's proper place on the Earth from a cosmocentric perspective.

SOLUTIONS TO FOOD AND HOUSING PROBLEMS

In his study of terraforming texts, Martyn Fogg argues that terraforming contains two subsets of planetary engineering: terraforming alien planets and terraforming Earth (90). Terraforming Earth includes such projects as land reclamation, genetic engineering for the improvement of agricultural production, and weather modification. *Alien Zone* and *Six Realms* both emphasize land reclamation projects on Earth. The two texts describe the use of advanced technology to explore various dimensions of both space and time in order to create more housing and food for an exponentially expanding human population. However, in contrast with Chinese terraforming narratives written between the 1950s and the 1980s, these two recent narratives express serious misgivings about humankind's excessive emphasis on aggressively reshaping nature to suit the unlimited growth of housing and agricultural production. At the end of both narratives, humans decide to destroy the new dimensions of space and new time zones that they had created, retreat to their original space, and heal the damaged Earth.

Written in 1999, *Alien Zone* is an apocalyptic fin de siècle novella. In the narrative, the population of Earth has already increased to fifty billion. Wars have broken out all over the world due to increasingly frenetic competition for food and other resources. In this crises-ridden time, Dr. Ximai (literally, "West Wheat") founds a farm called the Ximai Farm, which is managed by a computer and can produce enough foodstuffs to feed the world's entire population. This farm is located in a special zone where the passage of time is accelerated, thereby enabling a larger number of harvesting seasons than anywhere else on Earth. Initially, the pace of time's passing inside the farm is only two times faster than its pace in the outside world. Yet with skyrocketing increases in worldwide population and food consumption, the pace of time's passing within the farm speeds up to the point at which an entire year passing by inside the farm is equivalent to a mere ten minutes going by in the outside world. The farm is able to ship out its annual harvest every ten minutes, according to the world's general reckoning.

The technical defect of the Ximai Farm is that the acceleration of time is nonreversible. One day, the farm suddenly stops sending out food to the outside world. The protagonists Lin Chuan, Lan Yue, and their colleagues are dispatched to solve the problem. On the farm, these scientists discover that various creatures here have evolved at an accelerated speed in tandem with the accelerated growth of the crops. Among them, the intelligence of one type of creature has evolved nearly to the advanced level of human intelligence. The scientists call them "monsters" (*yaoshou* 妖兽) because of their giant bodies and advanced level of intelligence. These monsters sabotage the central computer system and try to take over and rule the farm. Though the scientists temporarily fix the damaged central computer and defeat the monsters, they know that the rapidly evolving monsters will eventually surpass human beings in overall ability and take over the human world. If this were to happen, human beings would either be ruled or wiped out by the monsters. At the end of the narrative, human beings decide to permanently shut down the farm and disconnect it with the outside world. The government initiates the *tuigeng huantian* (退耕还田) plan, which means they will restrict industrial development and convert the industry-used land to farmland. Lin Chuan and Lan Yue decide to return to the Ximai Farm and live there indefinitely to guard it for human beings. They and their offspring will evolve together with other creatures inside the farm. They hope that their high level of intelligence as humans will always surpass that of other organisms in the farm, and that the boundary between the Ximai Farm and the outside world will never be crossed. Upon the permanent shutdown of the farm, mass famine breaks out and the population plummets.

While scientists in *Alien Zone* modify the dimensions of time on Earth to meet the expanding needs of the skyrocketing human population, scientists in *Six Realms* modify the dimensions of space on Earth. In *Six Realms*, the government needs to open up more lebensraum for humans now that the global population has reached twenty billion. Physicists soon discover five other dimensions of space on Earth. These five dimensions are all arranged in accord with energy constants that differ from the Planck constant. Though these dimensions are not as stable as our own dimension that is in accord with the Planck constant, the government still decides to pursue development of these dimensions of space and implement mass migrations there. Eventually, there are six parallel dimensions of space on Earth to accommodate a total of seventy billion people. As time goes on, various problems emerge in all five of the new dimensions of space. For example, an avalanche in one dimension spills over into a desert in another dimension. This type of random overflow between different dimensions seems to increase in frequency, and leads to an increase in fatalities among human beings. At the end of the narrative, human beings initiate a *huigui jihua* (回归计划), or "plan to return home." Within 100 years, all human residents within the five unstable realms will have eventually moved back to the dimension which is in accord with the Planck constant—the world that nature has originally provided for human beings and all other living things. The other five realms will all be destroyed after the completion of the "plan to return home."

These two narratives reveal three pressing issues that human beings are facing in contemporary world. The first is the tension between a seemingly unlimited growth in population and Earth's carrying capacity. The second is the interplay between advanced technology and political power. The third is the idea of a feasible strategy to solve the problem of overpopulation.

Both narratives revisit some decades-old motifs in Chinese science fiction, such as overgrown plants and animals, radical land reclamation, and overly ambitious scientists. In *Alien Zone*, tree-size corn plants and gigantic animals remind readers of many PRC sf narratives written in the period from the mid-1950s to the early 1980s. In these earlier narratives, genetic engineering is used to increase agricultural productivity in both crops and livestock. Readers encounter elephant-size pigs and egg-size kernels of grain in Chi Shuchang's (迟书昌 1922–1997) short story "Elephants with Their Trunks Removed" (Gediao bizi de daxiang 割掉鼻子的大象, 1956) and Lu Ke's (鲁克b. 1924) "Grain Kernels as Big as Eggs" (Jidan ban da de guli 鸡蛋般大的谷粒, 1963). Readers can also come across rice paddies spreading endlessly like a golden carpet as far as the horizon—as well as apples, lychees, pears, and grapes all growing on the same tree in some short stories written in the late 1970s.[2] These Mao Era and early Reform Era PRC sf authors also envision using modern technology to drain submerged wetlands in the *Bohai Gulf* in order to reclaim land for agricultural uses such as growing legumes and herbs, along with constructing terraced dams in subterranean streams to generate hydro power and irrigate farm fields.[3] In most of these earlier PRC sf narratives, super-size plants and radical land reclamation projects are portrayed in entirely positive terms without the slightest hint of any downsides or unintended consequences. This early PRC sf fiction either echoes Mao Zedong's imperative that "Humanity must conquer nature," or chimes in with Deng Xiaoping's early Reform Era pronouncement that science and technology amount to advanced productive forces. The early Reform Era works also reflect Chinese writers' anxiety about the unsettling rapidity of their country's technological modernization after ten years of relative stagnation during the Cultural Revolution (1966–1976).[4] In these narratives, the scientist-protagonists are politically sound and professionally competent—both "red" and "expert," in Maoist terms. They see their botanical experiments as completing the mission that the CCP assigned them—as laying down a solid material foundation for their country's "socialist construction." The scientist-protagonists typically brim with the celebrated attributes of patriotism, a spirit of self-sacrifice, and optimism. The technical innovations or inventions described are the fruits of the scientists' collective aspirations and wisdom (Wu xxxvi).

He Xi makes a break from these earlier PRC science fiction narratives by infusing the time-worn motifs with strikingly contrasting implications. The overgrown plants, monsters, and giant reaping machines in *Alien Zone* represent excessive hubris in humankind's preoccupation with unlimited expansion and development. He Xi casts a critical view on the radical terraforming projects, and reflects on likely conflicts amidst the forces of technology, political power, and natural ecosystems. Science and technology come across as instruments for the fulfillment of humankind's endlessly increasing demands for more of everything. In *Alien Zone*, Professor Lan remarks, "What Dr. Ximai has done is no more than to comply with the needs of the populace" (He 346). In *Six Realms*, the antagonist Hao Nancun proclaims, "When we are short of energy, we discover atomic energy; when we are short of food, we discover the technology of genetic engineering; when we need more space, we discover new dimensions of space" (160). Another scientist named Jiang Zhexin comments, "We scientists have so easily deferred to the wishes of the populace, and have tried to satisfy them by any and all means" (171).

Science and technology not only have temporarily solved many of humankind's most pressing problems, but have also endowed some scientists with high-ranking political authority. In *Alien Zone*, Dr. Ximai wins election to the vice presidency of the government of the Earth Federation, and will likely win the presidency in the next election. Even though he knows the potential dangers that the Ximai Farm presents, he still single-mindedly decides to create a second farm of the same type: "He is no longer a scientist per se, but rather a politician" (337). Similarly, in *Six Realms*, Hao Nancun is not only a scientist, but also one of a very select few who can freely travel among the six different dimensions of space. Taking advantage of his exalted powers, he founds a cult called Free Paradise and declares himself a god in all six realms of space. Even though he knows that frequent natural disasters will occur in five of the realms due to the instability of their dimensions of space, he cunningly keeps the populace in the dark about this dangerous situation. Instead, Hao Nancun revels in the adulation from followers of his personality cult. These two scientists' devolution from scientist to political demagogue or even demigod implies that advanced technology might well become a convenient tool for political or theocratic autocracy.

In addition to such power-hungry scientist-protagonists, *Alien Zone* and *Six Realms* also portray some sober scientist-protagonists who hold cautious and critical views about high-tech terraforming of Earth. These scientists do not harbor the type of overly optimistic views about scientific and technological progress that were prevalent in Mao Era and early Reform Era PRC sf narratives. In *Alien Zone*, Professor Lan Jiangshui is the supervisor of Dr. Ximai. His theory of the "law of conservation of time and energy" provides theoretical foundation for the Ximai Farm. Professor Lan has foreseen the risk of implementing his theory in the real world. Therefore, he terminates this facet of his own research program, but cannot prevent Dr. Ximai from creating the farm. In the end, Professor Lan murders the Faustian Dr. Ximai in order to prevent him from establishing a second Ximai Farm. Similarly, in *Six Realms*, it is also a relatively sober scientist, He Xi (eponymous with the author), who murders the Faustian Hao Nancun and announces to the populace the severe hazards of the five new realms in space. Both narratives thus end with a seemingly justified murder of a Faustian scientist who has no scruples about the means used to attain an end.

The author's point seems to be that advanced science and technology should not be misused for the excessive and unsustainable exploitation of ecosystems. In that case, what is the author's apparent solution to the problem of human overpopulation, given Earth's limited carrying capacity? The author's solution is to reduce the human population and heal Earth's ecosystems. In *Alien Zone*, the population has already greatly exceeded Earth's carrying capacity through ordinary agricultural practices. Therefore, after the Ximai Farm is shut down, the deadliest famine in human history gets underway, resulting in a plummeting of the human population. This dramatic lowering of the population eventually enables a balance to be achieved in Earth's carrying capacity. Although the author does not directly refer to a massive famine in the companion narrative *Six Realms*, the reader can infer the strong likelihood of famine occurring on Earth. Once the residents of the five realms migrate back to the original realm that is aligned with the Planck constant, Earth will be once again become severely overpopulated. Famines and wars will thereupon likely reduce the human population to levels that are in accord with Earth's carrying capacity.

THE ETHICS OF TERRAFORMING

In addition to addressing the practical problems of shortages of food and space for humans on Earth, the two narratives also explore environmental ethics. I argue that terraforming projects and scientists' conflicting views about terraforming reveal three aspects of environmental ethics: the divergence between anthropocentric and cosmocentric viewpoints; humankind's proper place in an ecosystem; and human hubris.

With the emergence of terraforming fiction since the 1950s, the environmental ethics of planetary engineering or "geoengineering" have been discussed by philosophers, environmentalists, and scientists. Various scholars have contrasting views about whether human beings should transform a lifeless planet into a habitable planet. Such scholars as Erin Daly, Robert Frodeman, and Chris Pak have summarized various viewpoints in their studies of terraforming ethics. Arguments in favor of terraforming are normally articulated from an anthropocentric perspective. They argue that terraforming a lifeless planet would benefit human beings by "significantly advancing our scientific knowledge of the nature of life" or expanding the lebensraum for Earth's increasing population (Daly & Frodeman 145). Robert Zubin even regards the possible future terraforming of Mars as a demonstration that "the worlds of the heavens themselves are subject to human intelligent will" (Zubin 179). Don MacNiven explores the ethics of planetary engineering from three perspectives: the homocentric, the zoocentric, and the biocentric. On the basis of these viewpoints, MacNiven concludes that planetary engineering "would be morally permissible if either project helped protect and enhance the quality of terrestrial life" (MacNiven 442).

These views in favor of terraforming have two features in common. First, "they are all geocentric (Earth-bound) theories which automatically exclude from the moral universe Mars, the solar system and indeed, the universe as a whole" (MacNiven 442). Second, these viewpoints include within the moral universe nothing other than animate existence—human and other living organisms—and see them as intrinsically valuable because "life itself is the basis of value" (MacNiven 442). However, this Earth-bound perspective comes across as insufficient if we examine the moral issues of terraforming other inhabitable places in the universe such as extraterrestrial planets and moons.

Many scholars instead call for a cosmocentric perspective. Mark Lupisella and John Logsdon recommend a "cosmocentric ethic" for scrutinizing ethical matters related to terraforming (Lupisella & Logsdon 1). Daly and Frodeman capture the gist of a "cosmocentric ethic" as follows: "[It] places the universe at the center . . . [and] appeals to something characteristic of the universe (physical and/or metaphysical) which might . . . provide a justification of value, presumably intrinsic value, and allow for reasonably objective measurement of value" (qtd. Daly and Frodeman 140). If we extend our environmental perspectives from the anthropocentric and the geocentric to the cosmocentric, we may well develop a different approach to the ethics of planetary exploration. Similarly, MacNiven points out that cosmocentrism articulates a new ethical perspective that transcends the distinction between animate and inanimate entities: "Everything which exists has value" (MacNiven 442). This perspective requires us to assign the presumably inanimate planet of Mars with some sort of intrinsic value, such as its uniqueness and its role in the wilderness of space (442–43). In the light of cosmocentric ethics, philosophers such as Holmes Rolston and Alan Marshall argue that natural entities that lack living organisms also have intrinsic value, and that a planet bereft of life should "continue to exist in its natural state" (Daly & Frodeman 146-47).

In light of these contrasting views about the ethics of terraforming, we can detect an anthropocentric perspective in He Xi's two narratives. In both narratives, the artificial creation of more dimensions of space and time amounts to using "nature for the wellbeing of mankind" (MacNiven 441). Yet for the anthropocentric commentator, "The only question of terraforming or ecopoiesis is that whether the benefits to mankind would be great enough to warrant the costs of terraforming or ecopoiesis, and whether the resources required to accomplish these astonishing feats could not serve man in better ways" (441). In He Xi's stories, the cost that human beings must pay for creating these extra dimensions of time or space will eventually exceed the benefit they receive, and may even bring large-scale disaster down upon humankind. Therefore, sober-minded scientists must put an end to these terraforming projects by shutting down the new dimensions in time or space that reckless Faustian scientists had created.

Though the two narratives articulate anthropocentric views, they explore some profound moral issues—humankind's proper place in nature and our frequently arrogant attitude toward nature. In his study of the ethics of terraforming, Robert Sparrow points out two moral defects likely to be present when human beings engage in terraforming a planet: "an aesthetic insensitivity and the sin of hubris" (Sparrow 227). Sparrow argues that even though terraforming a lifeless part of space such as an extraterrestrial planet does not result in the suffering of sentient alien creatures, the action "reflects poorly on our character because it involves a blindness to beauty and an excessive faith in our own judgments and abilities" (239). In his view,

> Destroying the unique natural landscape of an entire planet to turn it to our own purposes reveals us to be vandals and brutes. It shows that we lead impoverished lives, being unable to respond appropriately to the beauty which is in the world (and on the world) around us. (233)

Sparrow's argument affirms that any lifeless area in space has its own intrinsic value; its natural beauty and its character of wilderness are independent of human beings' interests or benefits. In his view, any human attempt to "change whole planets to suit our ends is arrogant vandalism" (227). This arguably arrogant attitude reflects how we perceive our relationship with nature and our proper role in nature.

He Xi's narratives provide a sort of thought experiment for Sparrow's hypothesis about humankind's frequent arrogance. Human beings may well be playing God when they use advanced science and technology to shape the natural world to their ends. The author expresses his reflections on humanity's tendency to arrogance and the relation between human beings and nature through the mouths of some of his sober scientist-protagonists. In *Alien Zone*, Professor Lan confesses that he used to believe that science and technology could solve all of the problems that human beings face. According to Lan's previous viewpoint, the negative impact of technology on nature and society would be temporary, and would eventually fade into insignificance as science and technology advanced even further. However, his optimistic view has gradually changed over years. If the purpose of doing scientific research is to solve all the riddles of nature, he asks, why has Mother Nature gone to such great lengths to conceal the clues that might solve these riddles? It may well be that Mother Nature wants to prevent people from using technology to manipulate nature and destroy themselves (He 338). In *Six Realms*, the main character He Xi says:

> Humanity is so smart and is able to discover so many secrets of nature. If we liken Mother Nature to our own mother, then human beings are her clever and yet horrible children. These little kids are naughty and arrogant, and ceaselessly demands things from their mother. (180)

The confessions of both of these scientists reveal their awe for nature, as well as their apology for the way human beings have routinely viewed nature as ripe for plunder. Equipped with advanced science and technology, humankind has often arrogantly assumed that instead of living within the limits imposed by nature, humans can reverse this hierarchy and impose their own will upon nature without suffering any negative repercussions.

However, when humans seek to play God and willfully ignore the limits of our understanding, we will likely suffer the consequences of our arrogance and ignorance. In both narratives, terraforming the world incurs catastrophic failure. "The pride and fall go hand in hand" (Sparrow 237). In *Alien Zone*, when witnessing the prevalence of famine, Lin Chuan says: "This is the price that humankind must pay for its excessive exploitation of Nature" (He 347). He compares the monsters in the Ximai Farm and the mass famine to two bitter fruits that human beings have produced for themselves. "We have to swallow one or the other of them," he concludes (347).

When we examine the ethics of terraforming from a cosmocentric perspective, we need to reexamine "humanity's place in the cosmos" and "appropriate limits to human activities" (Sparrow 237). In environmental studies, scholars often differentiate the concepts of space, place, and environment. Lawrence Buell explains that "space is a broad term which connotes geometrical or topographical abstraction" (Buell 63). Place "can be as small as a corner of your kitchen or as big as the planet, now that we have the capacity to image Earth holistically" (67–68). Environment is something between space and place. It "refers to the complex surroundings we experience that can include both built and unbuilt nature" (Bernardo 3). Among these three different but related concepts, place is relatively personal, as we normally have an emotional attachment to it (3). In the same vein, Sparrow suggests the term "proper place" to indicate human beings' home in a spatial sense. In his view, "A proper place is one in which one can flourish without too much of a struggle" (Sparrow 238). He gives two reasons why a terraformed planet is not a proper place for us. First, when we terraform a planet, "The vast amount of effort required for us to sustain a presence there, even to the point of entirely transforming the planet, indicates that it is not a natural environment for us" (38). Second, a proper place is related to the idea of home. "This is a place which nurtures them, in which they grow up and reproduce, and which offers them some semblance of safety" (238). In the light of the idea of proper place, we can see that the zone of accelerated time and the five extra dimensions of space in the two narratives are not proper places for human beings. They are artificially created by technology, and are not self-sustainable. Our life has evolved in the dimension of space that is in accord with the Planck constant. Moving the surplus population to the five created realms in *Six Realms* is to "remove us from our proper place in relation to the natural world" (He 243). At the end of the narrative, the protagonist says: "The plan for returning home will be finished in a few minutes, and these five realms will be destroyed. These alien zones originally did not exist, and should not exist. They are just castles in the air" (181). The author compares the six parallel worlds to the six realms in Buddhism. This is how the novella got its title. One character in the novella says: "Buddha has prepared only one realm as a world for human beings, but we have racked our brains to overturn this arrangement" (171). Similarly, in *Alien Zone*, the sober scientists' permanent shutdown of the zone of accelerated time is a gesture of returning to our proper place in nature and on Earth.

CONCLUSION

Since the 1950s, numerous PRC science fiction narratives have dealt with the themes of terraforming the Earth. These themes are all closely related to the country's modernization drive, its economic takeoff, and the state-sponsored rhetoric of building a powerful China. The speculative projects in these narratives demonstrate humanity's power to manipulate nature and expand its footprint on Earth. The theme of terraforming is consistent with the discourse of development in modern China. In his study of the Chinese "new wave" science fiction, Mingwei Song indicates that "in Chinese sf the myth of development has been a constant focus, particularly since the early years of Mao's Republic" (92). The "new wave" Chinese sf writers continue to be preoccupied with development, while occasionally reflecting "on its ethical and technological effects" (Song 91). Such renowned writers as Han Song and Wang Jinkang have soberly reexamined "the myth of unlimited development and its disastrous effects" in their novellas *Subway* (*Ditie* 1991) and *The Reincarnated Giant* (*Zhuangsheng de jüren* 2006), respectively (Song 93). He Xi's two terraforming narratives discussed in this essay are in the same vein. Instead of celebrating the triumph of science and technology, the two texts demonstrate that advances in science and technology cannot fundamentally ease our shortages of food and housing. They endorse the message that "the overwhelming desire for development leads to uncontrollable results that will eventually collapse back on to the developers" (Song 93–94). The two narratives emphasize the themes of a return to nature and restraint in viewing technology as a silver bullet—returning to a proper place that nature originally provided, and restraint in applying technology too aggressively and without considering unintended consequences. It is only in this way that human beings can continue to exist in a sustainable fashion on Earth.

The two narratives are also textual sources for us to ponder the ethics of terraforming. They probe the moral question about whether humankind should terraform Earth to meet our needs. They open a space for us to reflect on humankind's frequently arrogant view of nature as ripe for plunder. Though both the implementation and termination of the geoengineering projects in the narratives are based purely on the perceived costs and benefits for human beings, these scenarios inspire readers to extend our perspective from human world to Earth as a whole, and even to the universe. Given the current state of global climate change, the depletion of natural resources, and environmental degradation, "We are required to respond to features of the world around us which are independent of our own interests" (Sparrow 235). While advanced science and technology enable us to transform aspects of nature within our reach, we should also exercise our sense of responsibility for maintaining the wellbeing of nature. We would do well to address and repair much of the damage our species has done to nature in the past. "Until we heal the Earth, we have no claim to any further space" (244).

NOTES

1. Six Realms was completed in October 1999 and was not published until 2002. This is indicated by the date of October 9, 1999, when He Xi wrote the postscript of *Six Realms*.

2. These images are presented in "A Golden Carpet Reaching to the Horizon" (Jin ditan pu shang tian le 1978) and "The Magic Orchard" (Mo yuan 1978) co-authored by Zhang Yiwu and Wei Liao.

3. These projects are depicted in Wang Guozhong's short story "Dragon in the Bohai Sea" (*Bohai jülong* 1963) and Liu Xingshi's "Eye of the Sea" (*Haiyan* 1979), respectively.

4. For detailed discussion of the Chinese terraforming narratives from 1950s to 1980s, see Hua Li "The Environment, Humankind, and Slow Violence in Chinese Science Fiction," 270–82.

WORKS CITED

Bernardo, Susan M. "Introduction." In *Environments in Science Fiction: Essays on Alternative Spaces*. Ed. Susan M. Bernardo. Jefferson: McFarland & Company, 2014. 1–7.

Buell, Lawrence. *The Future of Environmental Criticism: Environmental Crisis and Literary Imagination*. Oxford: Blackwell, 2005.

Daly, Erin Moore and Robert Frodeman. "Separated at Birth, Signs of Rapprochement: Environmental Ethics and Space Exploration." *Ethics and the Environment* 13.1 (Spring 2008): 135–51.

Ehrlich, Paul R. *The Population Bomb*. New York: Ballantine Books, 1968.

———— and Anne H. Ehrlich. *Population, Resources, Environment: Issues in Human Economy*. San Francisco: W. H. Freeman and Company, 1970.

Fogg, Martyn J. *Terraforming: Engineering Planetary Environments*. Warrendale, PA: Society of Automotive Engineers, 1995.

He, Xi. *Yi Yu* [Alien Zone]. *Liudao zhongsheng: He Xi kehuan zixuan ji* [Six Realms of Existence: self-selected sf works of He Xi]. Wuhan: Changjiang wenyi chuban-she, 2012. 323–49.

————. *Liudao zhongsheng* [Six Realms of Existence]. *Liudao zhongsheng: He Xi kehuan zixuan ji* [Six Realms of Existence: self-selected sf works of He Xi]. Wuhan: Changjiang wenyi chubanshe, 2012. 120–84.

Hua, Li. "The Environment, Humankind, and Slow Violence in Chinese Science Fiction." *Communication and the Public* 3.4 (2018): 270–82.

Lupisella, Mark and John Logsdon. "Do We Need a Cosmocentric Ethic?" Presentation at the International Astronautical Congress (Turin, Italy, 4 November 1997), IAA-97-IAA.9.2.09, International Astronautical Federation, 3–5 Rue Mario-Nikis, 75015 Paris, France.

MacNiven, Don. "Environmental Ethics and Planetary Engineering." *Journal of the British Interplanetary Society* 48 (1995): 441–43.

Pak, Chris. *Terraforming: Ecopolitical Transformations and Environmentalism in Science Fiction*. Liverpool: Liverpool University Press, 2016.

Song, Mingwei. "Variations on Utopia in Contemporary Chinese Science Fiction." *Science Fiction Studies* 40.1 (March 2013): 86–102.

Sparrow, Robert. "The Ethics of Terraforming." *Environmental Ethics* 21.3 (1999): 227–45.

Wu, Dingbo. "Looking Backward: An Introduction to Chinese Science Fiction." In *Science Fiction from China*. Eds. Patrick D. Murphy and Dingbo Wu. New York: Praeger, 1989. xi–xli.

Zubin, Robert. "The Case for Terraforming Mars." In *On to Mars: Colonizing a New World*. Eds. Robert Zubin and Frank Crossman. Ontario, CA: Collector's Guide Publishing, 2002. 179–80.

Chapter 6

Junkspace and Nonplace in Taiwan's New Eco-Literature

Peter I-min Huang

Taiwan is facing environmental problems that can be described broadly by resorting to globalization studies scholar Ulrich Bech's use of the term "global risk." A more accurate if also more polemical understanding of risk is postcolonial ecocriticism scholar Rob Nixon's "unequally distributed catastrophe." This chapter critically underscores the latter kind of risk, as it is playing out in East Asia, through the work of writers who question institutionalized policies and practices deleterious to planetary environmentally sustainable cultures and communities. In response to those policies and practices, scholars situated in the environmental humanities emphasize ecocritical arguments in literature and other kinds of cultural production that speak for preserving and caring for as opposed to replacing Earth's oldest environments. Such arguments distinguish the work of Ming-yi Wu (吳明益), Guangzhong Yu (余光中), Ming Xiang (向明), Qiao Zhong (鍾喬), and Hung Hung (鴻鴻), writers who are the main focus of this chapter. It highlights critical connections that Wu implicitly makes between indigenous rights and environmental rights in his novel *The Man with the Compound Eyes*, critiques of anthropogenic culture that Yu and Xiang make in several poems, and the eco-activist poetic messages of Zhong and Hung.

TAIWANESE URBAN ECOCRITICISM

East Asia has a long history of environmental literature, one that begins with the appearance of print and with, in narrower meanings, the modern environmental movement. In Taiwan, the beginning of that movement dates to the 1980s, to the postmartial law years, when environmental issues and indigenous issues, which are inseparable from environmental rights (if also

pitted against environmental rights in many instances), gained more visibility in public forums. In the following decades, Taiwanese literature scholars
began to focus on environmental arguments found in literary texts and in
other kinds of cultural and aesthetic production. This chapter represents that
scholarly cynosure. It does so toward the purpose of emphasizing the vital
importance of ecocriticism in Taiwan and elsewhere in East Asia, in a geological epoch, the Anthropocene, which is bearing apocalyptic witness to the
profound transformation of the environment by the human and pushing to
extinction spaces, places, and beliefs that embody ecocentric ways of living
with Earth.[1]

Catastrophic phenomena that associate with the Anthropocene and the
profound transformations of Earth by anthropogenic agencies probably have
not egregiously affected the majority of literature scholars in Taiwan. They,
like the author of this chapter, are likely to be relatively insulated and insured
against those phenomena as members of Taiwan's urban elite and upper
middle classes.[2] Of that community of scholars, nonetheless, those who spe-
cialize in and take an interest in ecocriticism and, more broadly speaking, the
environmental humanities, are acutely aware of environmental problems in
Taiwan that are the result of economic overreliance on and meretricious rationalization of the environmentally unsustainable nuclear power, coal power,
industrial agriculture, and electronics manufacture industries. For decades,
these industries have claimed almost sole rights to Taiwan, as they have in
other countries in East Asia. They have warmed this region of the world,
spewed out waste into its stratospheres and biospheres, and defended all of
that under the rhetoric of necessary and unavoidable costs.

As part of the work of critically appraising and teaching literature toward
the goal of bringing more attention to industries in Taiwan that downplay or
outright dismiss the environmental problems associated with those industries, ecocritics avail themselves of ecocritical theory generated in the West
as well as locally based and locally funded knowledge. Enlisting knowledge
from the formidable billage of literary theory and criticism generated out of
the West, many of those ecocritics, who include myself, attach themselves
to that literary theory and criticism like barnacles on a nineteenth-century
hull or microscopic plastics amassing in one of the floating garbage patches
that transformed the Pacific Ocean in the twentieth century. Many of us also
look closer to home, finding in East Asian knowledge bases arguments that
are germane to tackling environmental problems, while also recognizing that
using Western-based theory and criticism does not predictably or invariably
boil down to or equate with neo-colonial agendas. As postcolonial scholar
Rob Nixon defends the use of Western-based theory and criticism in the
context of the Western discourse of environmental justice, if in the East such
knowledge is "borrowed largely from the West," then it is being "blend[ed]"

with "local discursive traditions" and in its new "melded forms" being "adaptively redeployed as a strategic resource" (36).

The reliance by Taiwanese and other East Asian literary studies scholars on Western-based ecocritical theory has contributed discernibly to the awareness of the problems associated with global warming in Taiwan and the environmental unsustainable practices that have given birth to it.[3] What is dismaying is that ecocriticism in Taiwan, that which reflects either Western-based or Eastern-based environmental knowledge, still speaks for a relatively small number of literary studies scholars. That is profoundly disappointing given that levels of carbon dioxide have exceeded the four hundred parts per million thresholds here as well as elsewhere (Vaughan). Global warming is not slowing down. In Taiwan, other countries in East Asia, and around the globe, as a result of global warming caused by humans' environmentally unsustainable activities, Earth is experiencing tremendous losses. "Global warming," a catchall term for those losses, includes a wide range of causes and effects. For the first time in history, mass extinction of species due to human attitudes toward and direct violent intervention in other species is occurring. Ancient reserves and catchments of water (aquifers, snow caps, rivers, wetlands, and so forth) are dwindling rapidly. Industrial metals are building up in soils (where we grow our agricultural produce) and making those soils toxic to humans and other biological life. The Earth's rivers and oceans are filled with plastic refuse. There is more of the greenhouse gas of methane in the atmosphere as a result of landfills (a common source of waste disposal in Taiwan) and industrial agriculture. Dioxin—released, for example, when plastic waste is incinerated, the most common disposal method for treating rubbish in Taiwan as well as other countries in East Asia—and other "hyperobjects" are building up in the atmosphere.[4]

In Taiwan, literary studies scholars' efforts to address the human agent of global warming include two noteworthy studies: *East Asian Ecocriticisms: A Critical Reader,* edited by Simon C. Estok and Won-Chung Kim, and *Ecocriticism in Taiwan: Identity, Environment, and the Arts*, edited by Chia-ju Chang and Scott Slovic. Both collections include references to and discussions of two of Taiwan's most distinguished environmental writers, Ming-yi Wu and Kexiang Liu. Two other writers well worth reading for their role in challenging institutions and government bodies that either accelerate or do little to address global warming include the late Guangzhong Yu, one of Taiwan's (and China's) most revered poets; and Ming Xiang, a writer who, similar to Yu, has a significant critical reputation as well as a popular following in Taiwan.

In what remains of this chapter, I comment on several poems by Yu and Xiang that function as clarion calls to reimagine and reform human society such that it is not a burden to but rather a boon for the planet. I comment also on Wu's novel, *The Man with the Compound Eyes,* analyzing it for what it says about the friction between indigenous rights issues and superficial forms of environmentalism. Finally, I discuss the work of the activist-writers Qiao Zhong and Hung Hung. In that discussion I critically rely on a study entitled *Black Island: Two Years of Activism in Taiwan* by Michael J. Cole (寇謐將).[5]

Guangzhong Yu's "Visit to Flower Kingdoms" (花國之旅), appears in a collection entitled *The Sun Roll Calls* (太陽點名).[6] The target of Yu's poem is the organizers of the Taipei International Flora Exposition, also known as the 2010 Flora Expo, the first of its kind in Taiwan. After the exhibition closed in 2011, the facilities and grounds that were constructed and appropriated specifically for the exposition became the "Taipei Expo Park." The grounds and amenities were free to the public before the exhibition. After it opened, they were accessible for the price of a ticket. Today the "park" is a massive concrete site periodically lit up by festoons of heat- and carbon-dioxide emitting electric lights in the evening. The old darkness is gone, the old trees are gone, the old walkways are gone, and the green space that remains sits beside a huge concrete building. Yu's speaker comments on that privatization and sequestration as well as on the rhetoric of "international trade," "international travel," and "worldwide migration" that the organizers of the 2010 Flora Expo used to promote the exhibition and its main attraction, thousands of species of flowers flown or shipped into Taiwan. His speaker compares the highly ordered, controlled, and ostentatious exhibit to "a United Nations" of flowers, "an opera house" of symphonies, "a legal riot of colors," "a stream of consciousness of cubic geometry," a "palace of aestheticism," and a canvas by such European artists as Renaissance painter Botticelli, Impressionist painter Renoir, and Modernist painter Chagall. Yet, he or she is dismayed, even sickened, by the display.

Such sites as the 2010 Flora Expo have proliferated in the last sixty years. They identify the Anthropocene, an epoch defined by the human species and its anthropogenic creations. At the 2010 Flora Expo, flowers are "hand-cuffed" by plastic, pierced with wire, and propped up by agencies other than their own. Their final resting place is the garbage bin, "tomorrow's grave" (Yu 210–211), a descriptive that reminds me of ecocritic Timothy Morton's "junkspace" (*Ecology Without Nature* 85, 86, 91).

Junkspace refers to old city parks, squares, and plazas as well as other spaces in cities that corporate capital interests appropriate, privatize, commodify, and then abandon or do very little to rehabilitate when the sites are no longer economically viable. They include spaces ranging from apartment complexes and industrial park sites to carparks, storage sites, and exhibition sites. Junkspace abounds in both the East and the West. It epitomizes the Anthropocene and its "phantasmagoric carnival[s]" (Morton, *Ecology Without Nature* 86). It is capitalism's degraded spaces of infinitesimally slowly decomposing plastic, concrete, bitumen, Polyvinyl chloride, and stagnant contaminated water; it is the spaces that anthropocentric industrial capitalism "discards in its furious progress" (85).

Ming Xiang is the pen name of Dong Ping (董平), one of Taiwan's leading contemporary poets. Xiang migrated to Taiwan from China at the end of World War II. He subsequently enrolled in the Air Force Academy to study electronic communication and began to devote time to writing poetry. He became the editor of the *Blue Star Poetry Journal* (藍星詩刊) and went on to publish several anthologies of poetry, many poems of which now have been translated into several foreign languages (English, French, German, Hindi, Italian, and Japanese). Xiang still publishes to this day. In "Return from the Flora Exhibition" (花博歸來), Xiang's speaker, like Yu's speaker in "Visit to Flower Kingdoms," is both amazed and appalled by the 2010 Flora Expo, which he and thousands of other people have lined up for the length of several city blocks to visit. The numbers and types of flowers are dizzying. Many are hybridizations of anthropogenic and ecogenic matter, composed of plastic dyes and fixes as well as genetically modified plant tissue. If the exhibit is a "magnificent palace" and "paradise on earth," then it is also a place of "wasteful luxury." In "Wild blooming without a butterfly" (衆花怒舞，獨缺一枚蝶), another poem by Xiang about the 2010 Flora Expo, the speaker implicitly compares the exhibition to industrial agricultural sites in rural Taiwan. The flowers are detached and alienated from their own grounds, ecosystems, and environments. They "stand erect," they "dance," they are a "riot of color," but "why do they not attract a single butterfly?" The air itself in closed rooms of the exhibit space seems anthropogenic. It is "overwhelmed with too much human thought."

Xiang's and Yu's poetic ecocritical responses to the 2010 Flora Expo incisively question the substitution of anthropogenic cultures and spaces for ecogenic cultures and spaces by powerful industrial and commercial anthropocentric interests. They are manifestations of the Anthropocene, an epoch materially overdetermined by the human species and an epoch in which the environment now is, so to speak, more human than nonhuman. Xiang's and Yu's poems bring up the issue of who and what our governments and the corporate interests that support them spurn or neglect when they privatize, commercialize, and commodify Taiwan's oldest environments and old city parks that date to pre-industrial decades, when nature was not entirely disciplined and cordoned off by the human. Some of the given "who" are Taiwanese people who want to preserve Taiwan's oldest remaining environments. Some of the given "what" are lone surviving green spaces in cities. Some others of the given "what" are Taiwan's vastly diminished extant wetlands, unpolluted coastal waters, old-growth forests, butterfly species, leopard cats, and Formosan black bears.[7]

In asking questions about highly controlled privatized environments where both human populations and other species are monitored and have relatively little autonomy or freedom of expression and movement, Xiang's and Yu's poems bring to mind, further, what Julian Murphet, in an essay entitled "Postmodernism and space," calls "nonplaces" (118). They appear to offer freedom, or the "escape from necessity" (118), and many visitors do find a temporary kind of freedom in entering their gates. Nonetheless, the liberty that those spaces represent is more fabricated than real or sustaining. They are "a 'Disney [form of] realism'. . . sort of Utopian in nature, where [vested interests] carefully program out all the negative, unwanted element and program in the positive elements" and human bodies as consumers are highly regulated and controlled (Zukin qtd. in Murphet 118). A pertinent example is Taiwan's 2010 Flora Expo. Murphet uses the example of Disneyland in the United States. Writing about junkspace and referring to Dutch architect Rem Koolhaas's characterization of junkspace, Morton makes an argument that evoke's Murphet's critique of the "nonplaces" of theme parks such as the Taipai 2010 Flora Expo: "Trees are tortured, lawns cover human manipulations like thick pelts [and] sprinklers water according to mathematical timetables" (qtd. in Morton, *Ecology Without Nature* 86).[8]

The critical vocabulary of "junkspace" and "nonplaces" are useful for understanding Guangzhong Yu and Ming Xiang's implicit ecocritique of the transformation of an old Taipei city park into a highly commodified anthropogenic floral theme park. Other kinds of nonplaces and foredoomed junkspace include the ubiquitous shopping mall. Carolyn Merchant writes about the latter kinds of anthropogenic sites in an early seminal work of Marxist and feminist ecocriticism, *Reinventing Eden: The Fate of Nature in Western Culture* (an excerpt of which is included in the recently published anthology *Ecocriticism: the Essential Reader*, edited by Ken Hiltner).[9]

In *Reinventing Eden,* Merchant, one of the first scholars to forge critical connections between the work of feminists and environmental thinkers—links that have since come to be studied and taught under the area of inquiry of feminist ecocriticism—discusses the rise of the shopping mall, an example of what Murphy calls a nonplace. Such nonplaces are dystopic realizations of "the Garden of Eden" and "consumer capitalism's" vision of the "Recovery Narrative" of the biblical story of the fall of Adam and Eve (167–168). They purport to be "redemptive places of ecstasy," "places of light, hope, [and] promise," and "morally uplifting" (168). They offer "[s]anitized surroundings, central surveillance systems, noise restrictions," and "strict behavioral rules regulate the undesirable, homeless, and criminal elements of society" or else "keep out the socially undesirable" and "[reject] the naturally undesirable—weeds, pests, and garbage" (168–169). Their "artificial nature" exemplifies the control of "the human" over "the environment" or "nonhuman nature" (169). They are anthropogenic "gardens" surrounded by anthropo-genic "desert[s] of parking lots" and "the ideal of corporate postmodernism" (119, 167).

The nonplaces and junkspaces of industrialism and corporate capitalism have wreaked great destruction upon indigenous communities throughout the world, a subject that Ming-yi Wu writes about in his novel *The Man with the Compound Eyes*. Wu is well known within Taiwanese literature studies circles as well as in wider, global literary circles.[10] One of the first ecocritics to bring attention to Wu's novel *The Man with the Compound Eyes* was Serena Shiuhhuah Chou. In "Sense of Wilderness, Sense of Time: Mingyi Wu's Nature Writing and the Aesthetics of Change," Chou examined Wu's literary form as it reflects both a perdurable Chinese tradition of nature writing and a tradition that traces to the genre of "wilderness" nature writing in the West. Two other impor-tant Taiwanese ecocritics, Kathryn Yalan Chang and Rose Hsiu-li Juan, brought attention to Wu by drawing on studies published in the exciting new area of material ecocriticism and highlighting Wu's particular attention to plastic waste, namely the Great Pacific garbage patches. In "Taiwanese Mountain and River Literature from a Postcolonial Perspective," I also ecocritically situate Wu's novel in Taiwan's literary canons of mountain and river literature. In doing that, I also analyze the novel according to how it offers an counterdiscourse to mainstream anthropocentric discourse. The novel merits further attention for that counterdiscourse, as I hope to make clear here.

A fictionalized account of the long and ongoing history of genocide and dispossession of Aboriginal people in Taiwan, Wu's *The Man with the Compound Eyes* implicitly argues in effect that environmental rights are inseparable from indigenous rights. That argument is slowly gaining notice from middle-class Taiwanese. When one of the main characters, Havan, an Ami girl, is eleven months old, her mother Ina (which means mother in the language of the Ami) takes Havan with her to Taipei. They leave their home in Hualien County (on the east coast of Taiwan), the traditional place of the Ami, and relocate to the city in the north where, more so even than in their own country in Taiwan, to be Aboriginal is to be a subcaste of human. After many fruitless searches for a job, Ina finds work as a waitress in a karaoke bar. She entertains its male customers, drinking beer and eating peanuts with them as they sexually fondle and grope her body. She and Havan live with other Ami "migrants" in makeshift shanty houses that lie along the drybed of the river that runs through Taipei. It has long since used up by Taiwan's urban elite. Havan, now a young woman, learns that the government is plan-ning to evict the Ami people for the purpose of building an "environmen-tally friendly" bike path along the river, Da-feng, their chieftain, organizes a protest. In response, the government authorities cut off the electricity and water supply and tell the Ami that they need to relocate to an urban housing space. Ina refuses. She does not want to live "in a tiny pigeon hole" where one "cannot breathe" let alone "put up with the Han Chinese landlord's contempt and derision" (Wu 117). She and other Ami prefer to remain close to the river. It reminds them of "place," and more specifically of their own place in Taiwan and their river, Rita, in Hualian. They set up home there because other humans—namely, Taipei's wealthier urban populations—had spurned the area around the river. Now, those same populations want the river area back.

Havan has fallen in love with a boy whom she calls "Spider." Not long afterward, when the community learns that there will be heavy rain and possible flooding of the area where they live, Liao, Ina's boyfriend, a Han Chinese man from the city, persuades the reluctant Ina and Havan to leave the side of the river and take refuge in a downtown hotel. There, Ina and Havan watch the television coverage of the flood. The next day, Ina takes Havan back to their home despite the continued heavy rain. Many people have drowned or are missing. On the third day, when the flood waters still show no signs of subsiding, government rescue teams arrive. Havan joins them in the search for survivors and finds Spider's swollen body among the bodies of other. With the help of *Kawas,* the Ami's ancestor spirit, Ina finds the body of her boyfriend Liao.

Ina's belief in ancestor spirits brings up the subject of faith-based approaches to, influences upon, and the potentially significant contribution that spiritual beliefs can bring to, ecocriticism. Those approaches complement secular approaches and constitute a powerful counterdiscourse to dominant forms of worship of industrial development of the earth. Ecocritic Kate Rigby makes such an argument in the context of the imbalance in political dialogue in Australia between "Indigenous and non-Indigenous stakeholders" ("Spirits that Matter" 283). The imbalance reflects how much that dialogue is

> stymied by the prevalence of the modern Euro-Western onto-epistemology that infamously sunders nature from culture and construes the nonhuman world primarily as an object of knowledge, economic exploitation, or aesthetic appreciation, but rarely as agentic, communicative, and ethically considerable. (283)

What is needed instead is exchange whereby "contemporary forms of knowledge"—namely, knowledge generated out of the sciences and other secular areas of knowledge—are tempered by and "brought into conversation with nonmodern (and frequently non-Western) religions and philosophies" (289–290). Such exchange, then, includes bringing together "new materialist ecocritical thought"—represented by the work of Karen Barad and Serenella Iovino and Serpil Oppermann's edited anthology entitled *Material Ecocriticism*—and "older forms of nonreductive materialism, such as that which is articulated through Aboriginal narratives and practices of country" (284).[11]

Feminist ecocritic Greta Gaard makes a similar argument in "Mindful New Materialism: Buddhist Roots for Material Ecocriticism's Flourishing" in the context of Buddhist faith and its ancient concepts of impermanence (*anicca*), no-self (*anatta*), and dependent origination (*ipaticcasamuppāda*) (291). Buddhist tools and teachings, like ancient Australian Aboriginal knowledge, usefully complement relatively recent secular approaches to ecocriticism. Indeed, they have been there from the beginning. One can see them even in material ecocriticism, for in averring the "vast," immeasurable, and mysterious "interconnectedness of life-forms and [other agential] so-called nonliving matter," material ecocriticism indeed speaks for a spiritual component or "dimension" (292). Serena Shiuhhuah Chou makes that same argument in the context of her critique of Ursula Heise's term, "eco-cosmopolitanism": issues of "indigenous sacredness" and eco-spiritual forms of knowledge usefully complement and are highly relevant to understanding secular ecocritical understandings of "interconnectedness" (72).

As postcolonial ecocriticism scholars point out in studies of colonial histories that are similar to Taiwan's, in countless instances indigenous rights have been sacrificed under colonialist and neo-colonial industrial and corporate capitalist policies and practices.[12] The stakeholders of those policies and practices privatize and commodify green space where indigenous people have been living for thousands of years. Some of that green space includes areas that are now called "national parks," "nature reserves," "wilderness areas," and so forth.

Michael J. Cole, in a study entitled *Black Island: Two Years of Activism in Taiwan*, makes an implicit argument supporting that of postcolonial ecocriticism scholars in his study about the environmental movement in Taiwan. Indigenous people in Taiwan have long held up models for environmentally sustainable ways of living with their islands but have not been recognized let alone respected for providing those models.

The first environmental thinkers and activists in Taiwan who were given a political voice hailed from the elite classes (Cole 6). The environmentalists who followed them represented a more diverse cross-section of society and included grass-roots Taiwanese activists, individuals who in earlier decades would have been unable to speak; they included people working in nongovernment organizations, scholars, students, and individuals and groups working in several media organizations (6). Nonetheless, like the first wave of environmentalism, the second wave was dominated by the voices of Taiwanese of Han Chinese descent (inclusive of those who did not openly recognize or were not aware that they had indigenous ancestry or were of mixed ancestry). Those Taiwanese did not know or chose not to draw on indigenous knowledge because of the racist stigmas that attached to indigeneity up through the turn of the last century.

Despite the great strides in indigenous rights in recent decades, the environmental movement today tends to downplay ethnic differences, emphasizing instead Taiwanese people's shared democratic values, civic virtues, and support for social and environmental justice (Cole 8). The rationale for that downplaying is that without a strong, united front, the road to environmental sustainability will be longer and more arduous than it has already proved to be.[13]

Hung Hung and Qiao Zhong represent Taiwanese writers involved in Taiwan's environmental movement in recent decades.[14] Hung Hung is the pen name of Hung-ya Yen. Born in 1964 in Tainan County, Hung graduated from the Drama Department of Taipei National University of the Arts and became involved in activist writing. Since 2004, he has served as the organizer for Taipei Poetry Festival. Sheng Wu (吳晟), an environmental writer and activist who is as well known as Ming-yi Wu and Kexiang Liu are among East Asian ecocritics, wrote the preface for Hung's poetry anthology, *Plowing land on Jen-ai Road: Hung Hung Poems* (仁愛路犁田：鴻鴻詩集).[15] In his preface, he provides a brief summary of Hung's life and career, noting that Hung has attended every major anti-nuclear demonstration in Taiwan and describing Hung as being a writer who uses ordinary language powerfully to speak out frankly and vehemently against social and environmental injustice.[16]

In "God of the Fourth Nuclear Power Plant" (南無核四大神), Hung's speaker apostrophizes the poem's eponymous subject. He or she calls it "our omnipotent God." Its apparent "benevolence" reaches to "every corner of the island." Its "waste" provides livelihoods for Taiwanese people who are "used to being cheated" and so accept jobs in the nuclear industry. Other Taiwanese "worship [it] like devout pilgrims." In the last line of the poem, the speaker calls the industry an "evil spirit" and tells it to "retreat."

Taiwan has three operating nuclear power plants. In the early 1980s, Tai Power company officials began to dump large drums of nuclear waste on Orchid Island (Lanyu), situated about ninety kilometers south east of Taiwan. Representatives of Tai Power told the Aboriginal Tao residents that the drums were fish cans from a fish factory. They wanted to use Orchid Island as transit station and ship the nuclear waste stored there to waters near the Philippines. Following the signing of an international treaty in 1996 under which Taiwan could no longer dump nuclear waste in those waters, Tai Power could not carry out the second stage of its nuclear dumping plan. More than twenty thousand barrels of nuclear waste stayed on Orchid Island. In 2007, three decades after the first barrels were dumped at Orchid Island, government inspectors checked the site where the barrels were buried and found that many had rusted and were leaking their contents into the sea.

Despite protest from the Tao people, which included a large protest in 2013 in Taipei, the government has done next to nothing about the nuclear waste at Orchid Island.

Cole devotes a chapter in his book to the anti-nuclear protests in Taiwan, the most important events in the history of environmentalism in Taiwan (429–433). In the immediate wake of the Fukushima nuclear disaster in neighboring Japan, which occurred on March 11, 2010, public concern about nuclear power in Taiwan fired up again. The focus was on the fourth nuclear power plant, the intermittent construction of which has been ongoing for more than two decades. It is in Gongliao, a remote fishing village on the northeastern coast of Taiwan, about forty kilometers from the capital city of Taipei. Among the most outspoken protestors against the plant is Yi-hsiung Lin (林義雄), an activist with a long history tracing back to political persecution during the martial-law period.[17] In late April 2014, Lin staged a hunger strike. In support of it and the anti-nuclear cause as a whole, fifty thousand Taiwanese (of a total population of Taiwan of approximately twenty-four million people) protested in a march to the Presidential building in Taipei. In response, on April 24, then President Ying-jeou Ma announced the suspension of the construction of the fourth nuclear power plant (Cole 431). The "halting" of the construction of the plant is not a decisive victory for the anti-nuclear movement but does inspire those who do not want to see a fourth plant on their island to continue to demonstrate against nuclear power in Taiwan, which already has one of the highest number of nuclear plants per capita in the world.

Another major environmental concern for Taiwanese people refers to the unequal distribution of high industrial risk sites. On July 31, 2014, at midnight, Taiwanese were jolted out of their beds by news reports of a series of gas explosions in the underground pipelines in the southern industrial city of Kaohsiung. The explosions reached heights of fifteen stories and destroyed six kilometers of road. Thirty-two people died and more than three hundred people suffered injuries. Kaohsiung is the second largest city in Taiwan and the industrial capital of the south. Under Japanese colonial rule, the harbor city was designated to become an industrial zone, because it was situated far from the capital city of Taipei in the north, and a military base for the Japanese imperial army to launch southward campaigns to "conquer" Southeast Asia. When the KMT government took over control of Taiwan after the Japanese colonial government withdrew at the end of World War II, the new government continued to industrially develop the city, preferring it to Taipei for use as an industrial behemoth. The government concentrated petrochemical factories in and around the city, connecting them by a vast labyrinth of underground pipelines. In effect, it placed a particular population of Taiwanese people in a "sacrifice zone"; it chose a region of Taiwan whose human populations were stereotyped as being culturally inferior. Those populations identified with Taiwanese people of Han Chinese descent who migrated to Taiwan several hundred years before the Han Chinese people who came to Taiwan with Chiang Kai-Shek in the 1950s and became Taiwan's ruling elite.

The term "sacrifice zone" is postcolonial scholar Rob Nixon's (163). In the case of Taiwan's Kaohsiung residents, the government placed those Taiwanese populations in a sacrifice zone to serve the dubious "greater good" of the country (Nixon 163). The decision to base an industrial hub in a region where it was easier for the government to dismiss "the rights of the victims" because they were a "second class" of people critically ties to the work of postcolonial ecocriticism scholars. In the recent past, it was common for people from the north to ridicule people from the south for their "thick" Taiwanese accent when they spoke Mandarin and for being "country bumpkins" who chewed betel nuts and swore in the streets.

In situating a major industrial hub far from Taipei, the center of power, the colonial Japanese government and the subsequent Taiwanese government also practiced what Nixon calls "development as remote control" (169) and "the resource law of inverse proximity" (265). In other words, "the closer people live to the resources being 'developed,' the less likely they are to benefit from that 'development'" (Nixon 165). Nixon's particular critique also is useful to understanding why in Taiwan the government's proposal to build a fourth nuclear power plant has been on hold for many years. It is a site not far from Taipei, the bastion of Taiwan's political and industrial elite, the most environmentally minded of whom are strenuously objecting to the choice of the site.

In the case of Taiwan's massive petrochemical industry, environmentalists face an uphill battle equal to that which confronts anti-nuclear activists. It is as pervasive an entrenched as the nuclear power industry. Guangzhong Yu, in a poem entitled "Accusing a Chimney" (控訴一枝煙囪), compares the petrochemical industry in Kaohsiung to a reckless chain smoker whose exhalations drive sparrows away and trees to cough endlessly. It uses the entire city as an ashtray and blackens the lungs of Kaohsiung people. They become "dark and dying butterflies" crawling on the ground and unable to fly. "At Kaohsiung Station" (在打狗驛), by Kexiang Liu, is about a petrochemical disaster in 2014.[18] Another poem about the disaster is by Qiao Zhong, a poet, dramatist, and activist. In "Every Night, Kaohsiung, the Southern Harbor" (每一夜，高雄在南方的港都), Zhong speaks of the terrible inferno of July 31, 2014, calling it a harbinger of "every night, in Kaohsiung, the southern harbor."

Zhong, like Hung Hung, has been involved in environmental activism for many years in Taiwan. In "Anti-DuPont for thirty years" (反杜邦30年), Zhong relates his involvement in the protest in 1986 against the American corporation DuPont's plans for a titanium dioxide plant in Lukang, an old town in western central Taiwan. Zhong was serving in the army and noticed a tiny newspaper article about the opposition by people in Lukang, Zhong's hometown, to DuPont's plans. At the time, when he was a young man, Zhong knew little about issues of environmental protection, transnational pollution, and environmental pollution in the third world. When he began to learn about those issues after reading the newspaper article, his love for his hometown, his own local "place," spurred him to become actively involved in the move- ment against DuPont. That exposure included coming across the photographs by American photographer William Eugene Smith of the apocalypses in Minamata (Japan), Chernobyl (Russia), and Bhopal (India). To this day, Zhong remembers Smith's words, "I believe nothing but the wisdom inspired by deep emotions can redress the serious mistakes made in the world" (qtd. in Zhong).

The documentary journalist Shujuan Zhu (朱淑娟) writes about the anti-DuPont protests in Taiwan in a study entitled *The Road Not Taken* (走一條人少的路). The demonstrations turned over a new leaf in Taiwan's environmental movement. They were the first environmental as well as social demonstrations in Taiwan (Zhu 223). On March 8, 1997, after gathering at a local temple in honor of Mazu (the goddess who, in addition to the earth god and the goddess of compassion [Guanyin], is most worshiped by Taiwanese people) demonstrators led a protest march (Zhu 223). One hundred thousand people signed a petition against DuPont (223). Not long afterward, the com- pany withdrew its submission to build its plant.[19] In the same essay, Zhong quotes the theater critic Liang-ting Kuo (郭亮廷): "We need to re-historize the end of history. Return the power to those who have a face and a body. Let those patients who are dying not forgotten and invisible but remembered in history" (Kuo qtd. in Zhong, "Anti-DuPont for thirty years").

There is a weight of evidence of professional opinion for as well as grow- ing common and widespread knowledge of the fact that environmental crises are no longer something in the distant future. They are here and now. Powerful interests—namely, those represented by key stakeholders in the fossil fuel, agribusiness, coal power, and nuclear power industries—continue to meretriciously endorse and promote habits and attitudes that create and intensify environmental problems. In Taiwan, literature scholars specializing in ecocriticism and other disciplines falling under the environmental humani- ties are among those who are challenging business-as-usual, environmentally unsustainable practices. The texts that those scholars teach and study address such concerns as the dumping of nuclear waste in remote areas where a sig- nificant majority of the people are indigenous, the grip that the petrochemical industry has on the island, the nontransparent deals between the government and powerful multinational companies such as the deal between the govern- ment and Radio Corp of America (RCA), which was the taking advantage of cheap labor costs and easily circumventable environmental laws to fill people's bodies with high levels of carcinogens.[20] Those same texts, comprise a canon of Taiwanese literature that tells us that literature is not and never has been a site for articulating merely anthropocentric and speciesist beliefs. It is a site that can and is being used toward supporting and committing to environmental knowledge, environmental sustainability, and planetary healing.

NOTES

1. Ecologist Eugene Stoermer introduced the term in the early 1980s to refer to the increasing "transformative effects of human activities on the earth" (Haraway 49). It gained currency following the work of Paul Crutzen (Zalasiewicz et al., 14), who used it to refer to the impact of the human on the planet since the Industrial Revolution (approximately 1800 CE). Other scholars use the term to denote a new epoch following the Holocene Epoch but still in the Quaternary Period, forged by humans since the 1950s, and coincident with The Great Acceleration (Zalasiewicz et al., 14).

2. As stated in an MIT Press blurb for *Screen Ecologies: Art, Media, and the Environment in the Asia-Pacific Region* (by Larissa Hjorth, Sarah Pink, Kristen Sharp, and Linda Williams), the Asia-Pacific is "a key site of environmental degradation." It alerts us to the especial need for scholars in this region to engage with environmental issues as those appear inadvertently or overtly in literature and other aesthetic production.

3. The most well-known and compelling argument in defense of certain forms of globalization inclusive of the export of Western theory and criticism to the East is Ursula K. Heise's study, *Sense of Place, Sense of Planet.*

4. Timothy Morton uses the term "hyperobjects" to refer to a wide range of anthropogenic substances (*Ecology Without Nature* 130–135). They are highly toxic to humans and other animal species in small quantities, have been proliferating since the 1950s, and decompose very slowly, so they are building up in the biosphere. One of the most common hyperobjects is dioxin. Other hyperobjects are plutonium, used in the nuclear industry. Still others are chemical compounds in plastic goods such as polystyrene (the stuff of disposable plastic containers) and perfluorooctanoic acid (PFOA) (used in the manufacture of "nonstick" cooking pots and pans). As Nathaniel Rich writes about in an article entitled "The lawyer who became DuPont's worst nightmare," there are few restrictions on the production and disposal of many such "hyperobjects" because so little is known about them except in rare instances such as in the instance of the ongoing lawsuit against DuPont, which allegedly covered up for decades its knowledge about the harmful effects of PFOA.

5. Cole is the former editor of the *Taipei Times,* Taiwan's leading English language newspaper. He specialized in writing about cross-strait military issues. In December 2013, after almost eight years of editing work, he quit his job in order to become more involved in Taiwan's environmental movement (Cole 34).

6. The English translations of Chinese texts cited in this chapter are mine.

7. For the single most important study of the near extinction of the Formosan black bear, the last surviving largest terrestrial animal in Taiwan, see Mei-shui Huang's study entitled *In Search of Bears: Stories about me and Taiwan Black Bears,* For recent media coverage of leopard cats in Taiwan that survive in the wild, or in habitats where their movements are relatively unrestricted, see Ching-wen Chang's "Rescuing Leopard Cats"; "In Defense of the Leopard Cats in Taiwan" (no author); Coral Lee's "Beauty on the Prowl—Taiwan Leopard Cats"; and I-chia Lee's "Protesters rally over leopard cats habitat." For more on the ecocide of Taiwan's butterfly species, see Ralph.

8. Morton's discussion of "junkspace" in *Ecology Without Nature* is inspired by a so-titled essay by Koolhaas.

9. See Merchant, "Nature as Female."

10. Wu recently won the highly coveted United Daily Literary Award (聯合報文學大獎), and his latest novel, *A Stolen Bicycle,* is already receiving high praise from such literary critics as De-Wei Wang (王德威). As Wan-qian Chen (陳宛茜) writes about in a recent newspaper article, "Wu Ming-yi: novel changes my life," the novel is similar to *The Man with the Compound Eyes*, a critique of environmental imperialism in the form of a magic realism narrative.

11. See also Rigby, "Religion and Ecology."

12. See Huggan and Tiffin; DeLoughrey, Didur, and Carrigan; and DeLoughrey and Handley.

13. For a discussion of indigenous Taiwanese writers who address environmental issues see Huang (*Linda Hogan*).

14. For more on Kexiang Liu, who is as well known as Ming-yi Wu among East Asian ecocritics, see Yu-lin Lee's "Becoming-Animal: Liu Kexiang's Writing Apprenticeship on Birds." Liu earned an international reputation for his nonfiction, *A posthouse of migratory birds (Lü Niao de Yi Zhan)*. A fictional work, *He-lien-mo-mo the Humpback Whale*, also is one of Liu's most well-known and celebrated publications.

15. Sheng Wu is the author of one of the most important studies of the impact of the petrochemical industry in Taiwan. Entitled *Protecting Mother's River: Diary on the Jhuoshuei River* (守護母親之河: 筆記濁水溪) and published in 2014, the study is a revision of an early field study published in 2002, *A Notebook on Jhuoshuei River*. Between July 2001 and June 2002, when he was a writer in residence for Nantou County in central Taiwan, Sheng Wu traveled up and down the Jhuoshuei and recorded his experiences and reflections of that great river (2014, xii). This experience inspired a combination of a field trip report and a personal diary that led to the second study *Protecting Mother's River: Diary on the Jhuoshuei River* (2014, xii). For more on the poetry of Sheng Wu in comparative literature studies contexts, see two studies by Peter I-min Huang: "Corporate Globalization and the Resistance to It in Linda Hogan's *People of the Whale* and the Poetry of Sheng Wu" and *Linda Hogan and Contemporary Taiwanese Writers: An Ecocritical Study of Indigeneities and Environment.*

16. A poem by Hung that I do not comment on here but wish to note because it relates to my earlier discussion of nonplaces and shopping malls, is "Department Store" (百貨公司).

17. On July 15, 1987, the KMT government lifted the world's longest period of martial law (Zhu 223).

18. Kexiang Liu also protests against the concentration of petrochemical plants in Kaohsiung. For a detailed discussion of that, see Huang (*Linda Hogan* 137–138).

19. For further discussion of the anti-DuPont movement, see Reardon-Anderson.

20. See "Former RCA employees win decade-long legal battle."

WORKS CITED

Chang, Chia-ju, and Scott Slovic, eds. *Ecocriticism in Taiwan: Identity, Environment, And the Arts*, Lexington, 2016.

Chang, Ching-wen. "Rescuing Leopard Cats." *Business Weekly*, no. 1380, 28 April–4 May 2014, pp. 54–56.

Chang, Kathryn Yalan. "If Nature Had a Voice: A Material-Oriented Environmental Reading of *Fuyan ren* (The Man with the Compound Eyes)." *Ecocriticism in Taiwan: Identity, Environment, and the Arts*, edited by Chia-ju Chang and Slovic Slovic, Lexington, 2016, pp. 95–109.

Chen, Wan-qian (陳宛茜). "Wu Ming-yi: Novel Changes My Life." (吳明益：小說改變我人生). *United Daily* (聯合報), 28 June 2016, p. A11.

Chiu, Qiongyu (邱瓊玉), and Shihjie Chang (張世杰). "The Ruin Dome." (廢墟大巨蛋). *United Daily* (聯合報), 29 April 2016, p. A1.

Chou, Serena Shiuhhuah. "Claiming the Sacred: Indigenous Knowledge, Spiritual Ecology, and the Emergence of Eco-cosmopolitanism." *Cultura. International Journal of Philosophy of Culture and Axiology*, vol. 12, no. 1, 2015, pp. 71–84.

———. "Sense of Wilderness, Sense of Time: Mingyi Wu's Nature Writing and the Aesthetics of Change." *East Asian Ecocriticisms: A Critical Reader*, edited by Simon C. Estok and Won-Chung Kim, Palgrave Macmillan, 2013, pp. 145–163.

———. "Wu's *The Man with the Compound Eyes* and the Worlding of Environmental Literature." *CLCWeb: Comparative Literature and Culture*, vol. 16, no. 4, 2014, docs.lib.purdue.edu/clcweb/vol16/iss4/3.

Cole, Michael J. *Black Island: Two Years of Activism in Taiwan*. CreateSpace Independent Publishing Platform (On-Demand Publishing, LLC), 2015; *Black Island* (黑色島嶼), translated by Ming Li (李明), Yaxing Chen (陳雅馨), and Yuanyu Liu (劉燕玉), Business Weekly Publications (商周出版社), 2015.

DeLoughrey, Elizabeth, and George B. Handley, eds. *Postcolonial Ecologies: Literatures of the Environment*. Oxford University Press, 2011.

DeLoughrey, Elizabeth, Jill Didur, and Anthony Carrigan, eds. *Global Ecologies and the Environmental Humanities: Postcolonial Approaches*. Routledge, 2015.

Estok, Simon C., and Won-Chung Kim, eds. *East Asian Ecocriticisms: A Critical Reader*. Palgrave Macmillan, 2013.

"Former RCA Employees Win Decade-Long Legal Battle." *Taipei Times*, 18 April 2015, taipeitimes.com/News/front/archives/2015/04/18/200316192.

Gaard, Greta. "Mindful New Materialisms: Buddhist Roots for Material Ecocriticism's Flourishing." *Material Ecocriticism*, edited by Serenella, Iovino and Serpil Oppermann, Indiana University Press, 2014, pp. 291–300.

Haraway, Donna J. " Staying with the Trouble: Anthropocene, Capitalocene, Chthulucene."" *Anthropocene or Capitalocene?*, edited by Jason W. Moore, PM Press, 2016, pp. 34–76.

Heise, Ursula K. *Sense of Place and Sense of Planet*. Oxford University Press, 2008.

Hjorth, Larissa, Sarah Pink, Kristen Sharp, and Linda Williams. *Screen Ecologies: Art, Media, and the Environment in the Asia-Pacific Region*. The MIT Press, 2016.

Huang, Mei-shui. *In Search of Bears: Stories about me and Taiwan Black Bears* (尋熊記:我與台灣黑熊的故事). Yuan Liu Publishing Company (遠流出版社), 2012.

Huang, Peter I-min. "Corporate Globalization and the Resistance to It in Linda Hogan's *People of the Whale* and the Poetry of Sheng Wu." *East Asian Ecocriticisms: A Critical Reader*, edited by Simon C. Estok and Won-Chung Kim, Palgrave Macmillan, 2013, pp. 119–138.

———. *Linda Hogan and Contemporary Taiwanese Writers: An Ecocritical Study of Indigeneities and Environment*. Lexington, 2016.

———. "Taiwanese Mountain and River Literature from a Postcolonial Perspective." *Ecocriticism in Taiwan: Identity, Environment, and the Arts*, edited by Chia-ju Chang and Slovic Slovic, Lexington, 2016, pp. 29–39.

Huggan, Graham, and Helen Tiffin. *Postcolonial Ecocriticism: Literature, Animals, Environment*. Routledge, 2010.

Hung, Hung. "Department Store" (百貨公司). *Plowing Land on Jen-ai Road: Hung Hung Poems*, Preface by Sheng Wu, Black Eye Culture (黑眼睛文化), 2012, pp. 89–91.

———. "God of the Fourth Nuclear Power Plant." *Plowing Land on Jen-ai Road: Hung Hung Poems*, Preface by Sheng Wu, Black Eye Culture, 2012, pp. 56–58.

"In Defense of the Leopard Cats in Taiwan." *CNN iReport,* 16 April 2014, ireport.cnn .com/docs/DOC-1121311.

Iovino, Serenella, and Serpil Oppermann, eds. *Material Ecocriticism*. Indiana University Press, 2014.

Juan, Rose. Hsiu-li. "Imagining the Pacific Trash Vortex and the Spectacle of Environmental Disaster: Environmental Entanglement and Literary Engagement in Wu Ming-Yi's *Fuyan ren* (The Man with the Compound Eyes)." *Ecocriticism in Taiwan: Identity, Environment, and the Arts*, edited by Chia-ju Chang and Slovic Slovic, Lexington, 2016, pp. 79–93.

Koolhaas, Rem. "Junkspace." *October*, vol. 100, 2002, pp. 175–190.

Lee, Coral. "Beauty on the Prowl—Taiwan Leopard Cats." Translated by Scott Williams. *Taiwan Panorama*, 1 November 2012, taiwan-panorama.com/en/show_ issue.php?id=2012110111108E.TXT&table=2&h1=RWNvbG9neSBhbmQgRW5 2aXJvbm1lbnQ%3D&h2=QW5pbWFsIFByb3RlY3Rpb24%3D.

Lee, I-chia. "Protesters Rally Over Leopard Cats Habitat." *Taipei Times*, 17 April 2014, taipeitimes.com/News/front/archives/2014/04/17/2003588216.

Lee, Yu-lin. "Becoming-Animal: Liu Kexiang's Writing Apprenticeship on Birds." *Ecocriticism in Taiwan: Identity, Environment, and the Arts*, edited by Chia-ju Chang and Slovic Slovic, Lexington, 2016, pp. 155–162.

Liu, Kexiang. "At Kaohsiung Station" (在打狗驛). *China Times* (中國時報), 6 Aug. 2014, p. D4.

———. *A Posthouse of Migratory Birds* (*Lü Niao de Yi Zhan*) (旅鳥的驛站). Taiwan Environmental Protection Association (中華民國自然生態保育協會), 1993.

———. *He-lien-mo-mo the Humpback Whale* (座頭鯨赫連麼麼). Yuan Liu, 1999.

Merchant, Carolyn. "Nature as Female." *Ecocriticism: The Essential Reader*, edited by Ken Hiltner. Routledge, 2015, pp. 10–34.

———. *Reinventing Eden*. Routledge, 2003.

The MIT Press. Overview of *Screen Ecologies: Art, Media, and the Environment in the Asia-Pacific Region*. Edited by Larissa Hjorth, Sarah Pink, Kristen Sharp, and Linda Williams, MIT Press, 2016, mitpress.mit.edu/screen-ecologies.

Morton, Timothy. *Ecology without Nature: Rethinking Environmental Aesthetics*. Harvard University Press, 2007.

Murphet, Julian. "Postmodernism and Space." *The Cambridge Companion to Postmodernism*, edited by Steven Connor, Cambridge University Press, 2004, pp. 116–135.

Nixon, Rob. *Slow Violence and Environmentalism of the Poor*. Harvard University Press, 2011.

Ralph, Iris. "Local and Presentist Ecocritical Shakespeare in East Asia," *Tamkang Review*, vol. 47, no. 2, 2017, pp. 33–46.

Reardon-Anderson, James. *Pollution, Politics, and Foreign Investment in Taiwan*. Columbia University Press, 1992.

Rich, Nathaniel. "The Lawyer Who Became DuPont's Worst Nightmare." *New York Times*, 10 Jan. 2016, nytimes.com/2016/01/10/magazine/the-lawyer-who-became-duponts-worst-nightmare.

Rigby, Kate. "Religion and Ecology: Towards a Communion of Creatures." *Environmental Humanities: Voices from the Anthropocene*, edited by Serenella Iovino and Serpil Oppermann, Rowman & Littlefield, 2017, pp. 273–294.

———. "Spirits That Matter: Pathways toward a Rematerialization of Religion and Spirituality." *Material Ecocriticism,* edited by Serenella Iovino and Serpil Oppermann, Indiana University Press, 2014, pp. 283–290.

Schliephake, Christopher. *Urban Ecologies*. Lexington, 2015.

Vaughan, Adam. "Global Carbon Dioxide Levels Break 400ppm Milestone." *The Guardian*, 6 May 2016, theguardian.com/environment/2015/may/06/global-carbon-dioxide-levels-break-400ppm-milestone.

Wu, Ming-yi. *The Man with Compound Eyes*. Summer Publishing (遠足文化), 2011.

Xiang, Ming. "Return from the Flora Exhibition". *The Best Taiwanese Poetry, 2011* (2011 台灣詩選), edited by Jiao Tong (焦桐), Two Fishes (二魚文化), 2012, pp. 112–113.

———. "Wild Blooming Without a Butterfly." *The Best Taiwanese Poetry, 2011* (2011 台灣詩選), edited by Jiao Tong, Two Fishes, 2012, pp. 172–173.

Yu, Guangzhong. "Accusing a Chimney." *Yu Guangzhong's Humorous Poems* (余光中幽默詩選), edited by Xinghui Chen (陳幸蕙), Commonwealth Publishers (天下遠見出版社), 2008, pp. 226–227.

———. "Visit to Flower Kingdoms" (花國之旅). *The Sun Roll Calls* (太陽點名). Chiuko Publishers (九歌出版社), 2015, pp. 210–212.

Zalasiewicz, Jan, Mark Williams, and Colin N. Waters. "Anthropocene." *Keywords for Environmental Studies,* edited by Joni Adamson, William A. Gleason, and David N. Pellow, New York University Press, 2016, pp. 14–16.

Zhong, Qiao. "Anti-DuPont for Thirty Years". *United Daily*, 18 May 2016, p. D4.

———. "Every Night, Kaohsiung, the Southern Harbor" (每一夜，高雄在南方的港都). *China Times* 7 Aug. 2014, p. D4.

Zhu, Shujuan (朱淑娟). *The Road Not Taken.* Commonwealth Publishing (遠見天下文化), 2016.

Zukin, Sharon. *Landscapes of Power: From Detroit to Disney World.* University of California Press, 1991.

Part II

THE EMBODIED IMAGINARY

Chapter 7

The Loss of Genetic Diversity and Embodied Memories in Paolo Bacigalupi's *The Windup Girl*

Simon C. Estok and Young-Hyun Lee

Healing for the future means taking account of the damage that humanity has done and is continuing to do to the genetic diversity of the planet, a topic that Paolo Bacigalupi's *The Windup Girl* confronts very directly, with varying degrees of hope and despair. There is great reason to be hopeful about the future because the global genetic bank still has a lot in it, though we are draining it. Humanity began with a fortune, a lot of genetical material in the global bank to live with and share and enjoy, but it is diminishing, and we are the cause. The loss of genetic material means the loss of genetic diversity; the loss of genetic diversity means the loss of biodiversity; the loss of biodiversity means trouble. The fortune with which we began our tenure survives, but, we may rightly ask with Robert Frost, what to make of this diminished thing? E. O. Wilson has hauntingly warned that "We and the rest of life with us are in the middle of a bottleneck of rising population, shrinking resources, and disappearing species" (205), and it has become common knowledge that we could lose up to half of the species that remain on Earth by mid-century (see Gallucci). Half of our genetic inheritance being gone would be a disaster of inconceivable proportions and effects, a disaster from which there would be no recovery. To be hopeful about the future means to recognize the genetic inheritance that remains and to stop its loss at all costs: *The Windup Girl* charts the trajectory of one road to dealing with these complicated issues, a road that perhaps should not be taken—not least of all because of how the narrative remembers, perhaps unwittingly, racial and gender stereotypes.

This road not taken is studded with ethical mines and dilemmas, catastrophic ditches, and pools of tears. Emiko is the titular windup girl of Bacigalupi's novel. She is an engineered person, a kind of a sex slave, a genetically modified object of sexual traffic, a person of sorts—almost but not quite human. The narrator explains as much: "Look! She almost human!" (39).[1] Genetically engineering women to be like humans but who ultimately end up being only the property and tools of them (of men, in particular), the society that *The Windup Girl* describes is violently isolating. While the novel seems to do something radical and liberating at the end by giving Emiko agency and volition, it is important to consider that this volition is, by the logic of genetic materialism that the novel follows, hardwired. Critics have suggested, however, that Bacigalupi is doing something radical with race in this novel. Adam Trexler, for instance, notes that "too often, white, virile Americans are the heroes of science fiction novels, but *The Windup Girl* demands examination from other points of view" (212), and, indeed, though the novel begins with a focus on a white, virile, presumably American man, the narrative upsets our expectations through Emiko, the hero, if there is a hero here. But there are several problems with Trexler's recuperative reading: first, it does not properly address violence. This is odd because violence is a large part of what makes heroic white males heroic in fiction; the novel does not challenge this violence, and Emiko only comes into her own through her sudden almost psychotic violence (and it is precisely this violence that leads to the social and military crises in the novel); second, Emiko is clearly a racist and sexist stereotype; and third, the idea of enhanced women as a nexus for conversations about ethical conflicts within bioengineering is nothing new, and women's bodies and subjectivities have long been the conduit for conflict and conversation, long the materials causing and trophied by war. So even though Bacigalupi's novel is hardly radical, it does envision a future wherein genetic diversity is a central concern.

The lessons have been hard, and we still have not learned entirely well that genetic diversity is a matter of global concern. Memories of millennia upon millennia of the global biosphere are encoded in genetic material. Until humanity figures out that the hopes and promises of the future are entirely dependent on genetic materialism, there will be little hope for the future. Novels and film have recently registered recognition of the centrality of genetic materialism. These imagine the profound social and environmental implications stemming from the loss of the memories embodied in genetic diversity. Inextricable from matters as diverse as gender and food, formulations of memory in *The Windup Girl* reflect deep anxieties about loss of memory and the trauma that such loss entails. *The Windup Girl* depends on the recognition that it is well within human capabilities to exploit the plasticity of genetic material—for profit, for the benefit of humanity, for entertainment, and so on.[2] Not only is it well within our capacities; it is a necessary, truly unavoidable evil.

Genetic manipulation is the very reason that we maintain seed banks today. The purpose of the Svalbard Seed Vault is to maintain the availability of the genetic diversity of our crops—just in case. Just in case we need crops that can endure drought (or the genetic material that offers such endurance). Just in case we need crops that can endure frequent flooding (or the genetic material that offers such endurance). Just in case we need crops that can endure freezing and baking (or the genetic material that offers such endurance). Just in case. It is not simply a matter of enhancing crops and animals but also a matter of potentially bringing them back once they are gone. We learn early in *The Windup Girl* that extinct things can be resurrected: "Ngaw. It shouldn't exist. Yesterday, it didn't" (2). Not only plants are resurrected: "Felus domesticus was gone from the face of the world, replaced by a genetic string that bred true ninety-eight percent of the time" (30). One wonders about the other 2 percent! Even people are the objects of genetic intervention in the novel. Emiko and the other windups are the most immediate manifestation of genetically modified people in the novel, and one of the main issues that modifying people causes has to do with authenticity.

As a genetically modified person, the status of the windup is ambiguous. Whether they are indeed people or not is itself in question. Moreover, the question of free will (as if it were not already fraught enough before genetic intervention) becomes a central concern in the novel. The power of the gene is immense. E.O. Wilson notes that

> genes hold culture on a leash. The leash is very long, but inevitably values will be constrained in accordance with their effects on the human gene pool. The brain is a product of evolution. Human behavior—like the deepest capacities for emotional response which drive and guide it—is the circuitous technique by which human genetic material has been and will be kept intact. Morality has no other demonstrable ultimate function. (Wilson, *On Human Nature* 167)

Indeed, renowned science fiction writer Haruki Murakami has written in his dystopian epic *1Q84* that

> Human beings are ultimately nothing but carriers—passageways—for genes. They ride us into the ground like racehorses from generation to generation. Genes don't think about what constitutes good or evil. They don't care whether we're happy or unhappy. We're just means to an end for them. The only thing they think about is what is most efficient for them. (269)

It is a troubling thought. It is troubling in part because it questions the authenticity of our agency: is it real, or is it just in the service of the gene? If animals, including humans, are simply carriers, then it is not bodies that replicate but genes. Certainly, this is the position of Richard Dawkins: it is "the gene, the DNA molecule, [that] happens to be the replicating entity that prevails on our [. . .] planet," he explains in his prosaic masterpiece *The Selfish Gene* (248, 249). The gene is the final and irreducible unit of the bodies we prize and theorize, and indeed the Anthropocene discovery of the gene has profound implications for how we understand, conceptualize, theorize, use, inhabit, and value the body. At once throwing into doubt questions about agency and fantasies about the exceptionalism of human corporeality, current understandings of the gene really destabilize our confidence in "free will."[3] Emiko, understandably, therefore, has difficulty determining if it is her genes or her own will that determines her behaviors and if these behaviors help her to survive or not:

> she wonders if she has it backwards, if the part that struggles to maintain her illusions of self-respect is the part intent upon her destruction. If her body, this collection of cells and manipulated DNA—with its own stronger, more practical needs—is actually the survivor: the one with the will. (39)

Hiroko, another windup, at one point explains "It is in our genes. We seek to obey. To have others direct us. It is a necessity. As important as water for a fish. It is the water we swim in" (329). Ideas about genetic materialism and genetic determinism run throughout the novel, and for the windup to think that her behavior "is something in her genetics . . . deeply ingrained" (360) is uncontentious within the narrative world of the novel.

One of the strengths of this novel is that it subtly understates the dangers of genetic technologies while at the same time recognizing the importance and inevitability of gene manipulation. Siddhartha Mukherjee explains that the gene is "one of the most powerful and dangerous ideas in the history of science," and that it is "the fundamental unit of heredity, and the basic unit of all biological information" (9). He compares it to the other two fundamental units discovered during the twentieth century (the atom and the byte) for its potential danger, and for each, the potential dangers are in moving from simply trying to understand toward trying to exploit the complexities of these basic units. Mukherjee explains that "this transition—from explanation to manipulation—is precisely what makes the field of genetics resonate far beyond the realms of science. It is one thing to try to understand how genes influence human identity or sexuality or temperament. It is quite another thing to imagine altering identity or sexuality or behavior by altering genes" (11–12). Altering memories is a dangerous game, since memories (genetic and social) protect not only the past but, more importantly, the future.

Where the future leads depends enormously on who controls information, data, memories. It is for this reason that the battle for control over data and facts is so important. In the United States today, there is definitely something rotten in the state of the Union. When *Time* magazine is forced to run an issue with "Is Truth Dead?" emblazoned on its cover, we know that we have entered a new age, one in which opinions vie for supremacy against verifiable facts in the public understanding of truth. In one of the *Time* articles, Nancy Gibbs explains that "For Donald Trump, shamelessness is not just a strength, it's a strategy. . . Whether it's the size of his inaugural crowds or voter fraud or NATO funding or the claim that he was wiretapped, Trump says a great many things that are demonstrably false" (Gibbs), an average of "15 false claims a day in 2018," according to one CNN report (see Kessler). By 2020, things have clearly spiraled out of control with this president and his dictatorial control and manipulation of data, often resulting in outright lies. While one doesn't want to encourage censorship, at the same time there is clearly a question about the need to discourage falsehoods. Amazon appears to have gotten this message. The recent and continuing outbreak of measles in the United States seems to have prompted some change at Amazon: witness a recent headline running "Anti-vaccine movies disappear from Amazon after CNN Business report" on CNN.com (see Sarlin). It would have been better to have not carried the movies in the first place. The work of Dr. Andrew Wakefield connecting vaccinations with autism has long been known to have "been a fraud and a hoax" (McIntyre 83). Eleven people have died of measles in the United States since 2000. Hundreds of thousands more have died of COVID-19, deaths which have been the direct result of a very dishonest US President. But the working of media data pales in its importance to the control and manipulation of genetic material and the information it carries—information central to everything that happens in *The Windup Girl*.

At one point, a character by the name of Hock Seng is tempted to steal genetic material in order to regain the position that he has lost:

> The blueprints are there. Just inches away. He has seen them laid out. The DNA samples of the genehacked algae, their genome maps on solid state data cubes. The specifications for growing and processing the resulting skim into lubricants and powder. The necessary tempering requirements for the kink-spring filament to accept the new coatings. A next generation of energy storage sits within his grasp. And with it, a hope of resurrection for himself and his clan. (36)

What is remarkable here is the nonchalance: genes and genetic technologies are just resources for self-advancement. The effects of these technologies are not the issue. Indeed, though pithy at times, the narrative is accurate to note that "Politics is ugly. Never doubt what small men will do for great power" (306).

The global effects of what small men do with genetics can be immense, and much of this has to do with how genetic memory is refashioned. As we write this chapter, the COVID-19 global pandemic is humbling humanity, reminding us both of our mortality and our connection with the natural world. And if there is one thing that we have learned from the pandemic, it is that human behaviors are the direct cause of many of the viral agents and diseases that humanity has faced and will face—from swine flu to bird flu, from mad cow disease to COVID-19, the pathogens have grown out of food production practices that involve animals. We'd be better off vegetarian. Another thing that we have learned, however, is that our research labs throughout the world house pathogens that could, theoretically, be released at any time—an idea that seems to have fascinated President Trump enough to accuse China of intentionally releasing the virus that caused COVID-19. In *The Windup Girl*, such pathogens are weaponized, Anderson Lake musing at one point that he is "inoculated against diseases that haven't even been released yet" (363). In many ways, *The Windup Girl* is about grappling with various threats by manipulating genetic material. These technologies are responses to huge losses of global genetic diversity. These are losses of embodied memory. Memory thus must be understood broadly in this novel both as something that people recollect and as data that genes carry, data systems that are manipulable. Yet, while genetic science seems to be advanced in twenty-third-century Thailand, it is apparently not an exact science, and things pop up (such as Emiko's military prowess, hidden, apparently, throughout much of the novel).

Obviously, one of the effects of genetic modifications in our own world and in the world of *The Windup Girl* is that it becomes impossible to extricate genetic pollution from the global gene pool.[4] At what point does it become impossible to retrieve the wild? The devastation due to GMOs can be exponential, with mutations and variations being entirely unmanageable—quickly resulting in the loss of the memories that have been coded into wild genes over millennia, memories cut out and lost in modified genes. Lake complains that "Blister rust is mutating every three seasons now" (6). This is a book that is explicitly about genetic tampering and about the dangers of it. There are "recreational generippers [who] are hacking into . . . [genetic] designs" (6); there are companies whose job is to remake foods and species; there are, all the while, memories being sacrificed in ways that are simply not known.

In addition to raising questions about the many variable effects of genetic manipulation that are simply unknown (and perhaps unknowable), we have here the paradox of what we might call the residual agency of genetic material. What traits will become emergent and what will recede, and how do genetic strings and codings determine these? This is fire, and the characters in *The Windup Girl* have been playing with it for quite some time. It is a warning for us: do we really want to go down that long and winding road?

Certainly, genetics offer great hope for humanity; however, genetic technologies also pose great dangers. The prospect of genetic manipulation of human beings has long been a troubling and ethically fraught topic. Whether it is designer babies in China[5] or Nazi eugenics in Germany, whether it is Svalbard seed vaults in Norway or genetic engineering of COVID-19 vaccines, genes have our attention—and rightly. Attention to genetic materialism is essential for the future, for better or worse, for understanding losses and potential gains, and for grasping the enormity of memory that resides in genomic material—and memory once lost is potentially lost forever. There is good reason to hope for the future because writers of fiction and nonfiction alike have not shied away from these topics. Yet, the revulsion toward such research has been powerful, a revulsion that Jonathan Gottschall references in his part of the introduction to *The Literary Animal: Evolution and the Nature of Narrative*:

> I quickly learned that when I spoke of human behavior, psychology, and culture in evolutionary terms, their [other professors and graduate students] minds churned through an instant and unconscious process of translation, and they heard "Hitler," "Galton," "Spencer," "IQ differences," "holocaust," "racial phrenology," "forced sterilization," "genetic determinism," "Darwinian fundamentalism," and "disciplinary imperialism." (xx)

Although the work of scientists in gene theory, evolutionary biology, and cognitive neurology has used ideas about genetic determinism in nefarious ways throughout history, gene theory is not intrinsically ethically compromised. The work that has been done initiating such research, however, has been understandably defensive and tentative:

> It is not clear why Darwin—whose enduring impact on knowledge and politics is at least as strong as that of Hegel, Marx, or Freud—has been left out of feminist readings. It is perhaps time that feminist theorists begin to address with some rigor and depth the usefulness and value of his work in rendering our conceptions of social, cultural, political, and sexual life more complex, more open to questions of materiality and biological organization, more nuanced in terms of understanding both the internal and external constraints on behavior as well as the impetus to new and creative activities. (Grosz 24)

The revulsion toward integrating biological theories with matters of culture, the arts, and so on is indeed well known.[6] Paulo Bacigalupi offers a potent warning about this matter. Scientists imagine possibilities, to be sure, but so does literature—possibilities that we should sometimes avoid as well as ones that we would do well to embrace. The toxic chemicals from the kinkspring factory in *The Windup Girl* have profound immediate corporeal implications (the increase of prostate, ovarian, and breast cancers) and longer-term genetic consequences, helping us to imagine real possibilities that could result from our current behaviors, possibilities that disrupt the memory ingrained in our genes (that have evolved through billions of years). Imagining these possibilities can help us to avoid them and preserve those memories—and the scale of loss and of potential loss is staggering: "So many . . . things are simply gone" (72) in the novel. The realities of the world outside of the novel are equally dizzying.

For Scott Selisker, scale is centrally important in *The Windup Girl*, and he sees Bacigalupi as resolving some of the problems of scale through his use of GMOs: "*The Windup Girl* inventively enables readers to relate to the very small spatial scales and the long temporal scales at which the genome and its effects are most visible" (518). We see through the novel our potential separation from our past and out from our memories. We mourn our separation from pasts (and their geographies), pasts to which we cannot return. This nostalgia and "solastalgia"—a "place-based distress in the face of the lived experience of profound environment change" (Albrecht 12)—constitute what E. Ann Kaplan describes as "pretrauma," a condition in which "people unconsciously suffer from an immobilizing anticipatory anxiety about the future" (xix). Andrew Hageman usefully explains that

> the novel compels the reader to approach the extreme difficulties involved in imagining how and why the concepts of "Nature," the "human being," and ecologically sustainable capitalism must be disassembled to discontinue the indefinite reproduction of ecologically devastating attitudes and actions. (Hageman 284)

The characters themselves are well aware of nostalgia and loss:

> Gibbons snorts. "The ecosystem unravelled when man first went a-seafaring. When we first lit fires on the broad savanna of Africa. We have only accelerated the phenomenon. The food web which you talk about is nostalgia, nothing more." (264)

But it is something more: food webs evolved with humanity, and to suddenly break these webs or yank them away is traumatizing. One of the problems we face today is that the rapid changes to food production that began following the Industrial Revolution intensified our separation from the past, from our ways of knowing and producing our foods. The McDonaldization of global food webs is the real acceleration of "the phenomenon" to which Gibbons refers. And as many scientists note, today "food production is among the largest drivers of global environmental change by contributing to climate change, biodiversity loss, freshwater use, interference with the global nitrogen and phosphorus cycles, and land-system change (and chemical pollution)" (Willet et al 447). They also note that "a revolutionary change in food systems to support human health and environmental sustainability is essential" (451) and that "human appropriation of land for food production, is the greatest driver of biodiversity loss" (467). *The Windup Girl* shows the trajectory of this biodiversity loss. In the world Bacigalupi describes, however, like the world of Jurassic Park, separation from the past through extinctions is negotiable—indeed, negotiable in ways that are only possible in sci-fi, a point Hageman makes well:

> as a genre science fiction is well positioned to contribute to social conversations about the role of technology in possible ecological futures . . . within science fiction, as well as literature more broadly, *The Windup Girl* represents a new and sophisticated engagement with ecology, technology, and geopolitics. What differentiates Bacigalupi's novel is that it approaches key concepts underwriting the ecological crises in the novel and outside of it with an uncompromising speculative vision that brings their inherent contradictions distinctly into view. Among these concepts, one of the most fundamental is the belief that capitalism can readily be retrofitted into a sustainable economic system with greater profit margins than the current version. The narrative brings to light structural contradictions within this concept. Specifically, Bacigalupi remaps global capitalist geopolitics, interrogates the future of the nation-state in the face of transnational corporations, and explores the dynamic between capitalist commerce and ecological sustainability. (Hageman 284)

Even so, as we read the novel, our fear of what is to come is visceral. We have a lot to lose, and Kaplan's pretrauma is very real. For Kaplan, "future time is a major theme, along with thinking through the meanings and cultural work (including that pertaining to race and gender) that dystopian pretrauma imaginaries perform in our newly terrorized historical era" (4). It is an unrelenting future that Bacigalupi offers, one that says as much about current Western anxieties toward Asia as they do about our current climate trajectories—after all, the novel offers the militarization and sexual objectification of a genetically engineered Japanese woman in a Southeast Asian country (Thailand), not a genetically engineered white man in Alaska.[7]

It is perhaps the racial issues of this novel that are most problematical. Jungyoun Kim has argued persuasively about how "the politics of representation in Bacigalupi's novel are suspect in their construction of Orientalist and sexist stereotypes as a frame" (566): Kim explains that "Bacigalupi's use of an 'exotic' setting discloses the Orientalist perspective . . . [and that] a particular Asian tradition, that of the *geisha*, is appropriated in the image of the windup girl, revealing Western fantasies of Asian women" (569). Kim goes on at considerable length describing the relentless exoticizing of Asia and the essentializing of "the East" in *The Windup Girl* and situates the novel within a context that is clearly anti-Asian and Asiaphobic:

> [its] depictions of Thais, Chinese, and Japanese are consistently based on historically, politically, and socially prevalent Western images of the Orient. The narrative writes Thailand as politically corrupt, sexually promiscuous, and entrepreneurially "incompetent" (14), while Thais "simply lack the spirit of entrepreneurship" (132) . . . [Bacigalupi] emphasizes . . . stereotypes of racial and ethnic prejudice. (571)

Kim tangles with questions about the racial and gender politics of representation and the Orientalist and sexist stereotypes that seem to be involved in the novel's critique of genetic modification and food transformation technologies. At least part of the argument here is that it matters who speaks for whom and that although Bacigalupi effectively exposes the dangers of mass produced genetically modified food and the transformation of food into industrial products by Western-based multinational agri-corporations, setting the novel in Asia and reproducing Western fantasies and stereotypes of Asia seem to run counter to the overall thematic thrust of the novel. While the novel does, as Kim argues, succeed in showing how "the dangerous and uncertain future of food transformations is entangled with racial and gender matters in the novel," the novel is in some important ways compromised by the stereotypes it deploys. The monstrosities that line up along axes of racial difference in this novel compel the reader to wonder why this novel is set in Asia. What exactly is going on ideologically here? There is no doubt the memories of Western xenophobia and racism imbue this novel. "The monster dwells at the gates of difference," Jeffrey Jerome Cohen notes (7). Throughout history, there have indeed been many indices offering imagined boundaries of the human (sexuality, race, gender, diet, deformity, disability, behavior, and so on), and these become pliable via genetic manipulation in *The Windup Girl*—but the novel challenges neither the sexism nor the racism of its own narrative. Bacigalupi offers many clear warnings, but it is perhaps the narrative's seemingly unwitting reinscribing of the current trajectories of race and gender within the context of cli-fi issues that is most important: part of changing our ethical relationship with nature involves changing how we relate with each other—a point that the novel just does not seem to grasp.

NOTES

1. The remainder of this paragraph appears in a slightly different form in Estok, "Detachment and Division." See also Homi Bhabha's discussion of colonial

subjectivity, of the "recognizable Other, *as a subject of a difference that is almost the same, but not quite*" (86, *emphasis* in original).

2. Bacigalupi is not the first to fictionalize this kind of topic: Michael Crichton's *Jurassic Park* brought back dinosaurs to the thrill and horror of readers as early as 1990, and three years later Steven Spielberg brought this novel to a much larger constituency through what would become a blockbuster film and, ultimately, a franchise of sequels.

3. This and the previous two sentences appear in a slightly different form in Estok, "Introduction" 2.

4. Genetic pollution is an obviously problematical and controversial term that describes the diffusion of humanly modified genes into natural systems via avenues such as cross-pollination, inter-breeding between domestic and wild populations, and intentional intervention. It is a topic that because of space constraints we cannot address here.

5. The Chinese scientist "who stunned the world" in 2018 by producing "genetically edited babies, has been found guilty of conducting 'illegal medical practices' and sentenced to 3 years in prison" (Normile).

6. Parts of the preceding two paragraphs appear in Estok, *The Ecophobia Hypothesis* 26–27.

7. This paragraph and the next appear in a slightly different form in Estok, "Detachment and Division."

WORKS CITED

Albrecht, Glenn. "'Solastalgia': A New Concept in Health and Identity." *Philosophy and Nature*, vol. 3, 2005, pp. 45–50.

Bacigalupi, Paolo. *The Windup Girl*. Nightshade, 2009.

Bhabha, Homi. "Of Mimicry and Man: The Ambivalence of Colonial Discourse." *Discipleship: A Special Issue on Psychoanalysis*, vol. 28, Spring, 1984, pp. 125–133.

Cohen, Jeffrey Jerome. "Monster Culture (Seven Theses)." *Monster Theory: Reading Culture*, edited by Jeffrey Jerome Cohen. University of Minnesota Press, 1996, pp. 3–25.

Dawkins, Richard. *The Selfish Gene*, 40th anniversary. Oxford University Press, 2016.

Estok, Simon C. "Detachment and Division: Militarization, Geography, and Gender in The Windup Girl." *Mushroom Clouds: Ecological Approaches to Militarization and the Environment in East Asia*, edited by Simon C. Estok, Iping Liang, and Shinji Iwamasa. Routledge, pp. 141–156.

———. *The Ecophobia Hypothesis*. Routledge, 2018. Rpt with revisions and corrections, 2020.

———. "Introduction to Special Cluster on 'The Body and the Anthropocene.'" *Neohelicon: Acta comparationis litterarum universarum*, vol. 47, 2020, pp. 1–7.

Gallucci, Maria. "Half of the World's Species Could Become Extinct, Biologists Say." *Mashable*, 26 February 2017. https://mashable.com/2017/02/26/biological-di versity-vatican-conference/?utm_cid=hp-r-20#inDsUMaICkqO.

Gibbs, Nancy. "When a President Can't Be Taken at His Word." *Time*, 22 March 2017. http://time.com/4710615/donald-trump-truth-falsehoods/.

Gottschall, Jonathan, and David Sloan Wilson. "Introduction: Literature—A Last Frontier in Human Evolutionary Studies." *The Literary Animal: Evolution and the Nature of Narrative*, edited by Jonathan Gottschall and David Sloan Wilson, Northwestern University Press, 2005, pp. xvii–xxvi.

Grosz, Elizabeth. "Darwin and Feminism: Preliminary Investigations for a Possible Alliance." *Material Feminisms*, edited by Stacy Alaimo and Susan Hekman. Indiana University Press, 2008, pp. 23–51.

Hageman, Andrew. "The Challenge of Imagining Ecological Futures: Paolo Bacigalupi's *The Windup Girl*." *Science Fiction Studies*, vol. 39, 2012, pp. 283–303.

Kaplan, E. Ann. *Climate Trauma: Foreseeing the Future in Dystopian Film and Fiction*. Rutgers University Press, 2016.

Kessler, Glenn. "A Year of Unprecedented Deception: Trump Averaged 15 False Claims a Day in 2018." *The Washington Post*, 30 December 2018. https://www.washingtonpost.com/politics/2018/12/30/year-unprecedented-deception-trump-averaged-false-claims-day/? noredirect¼on&utm_term¼.15c7229c9e68.

Kim, Jungyoun. "The Problematic Representations of the Orient, Women, and Food Transformations In Paolo Bacigalupi's *The Windup Girl*." *Kritika Kultura*, vols. 33–34, 2019/2020, pp. 570–583.

McIntyre, Lee. *Post-Truth*. The MIT Press, 2018.

Murakami, Haruki. *1Q84: A Novel*. Alfred Knopf, 2011.

Mukherjee, Siddhartha. *The Gene: An Intimate History*. Scribner, 2016.

Normile, Dennis. "Chinese Scientist Who Produced Genetically Altered Babies Sentenced to 3 Years in Jail." *Science*, 30 December 2019. https://www.sciencemag.org/news/2019/12/chinese-scientist-who-produced-genetically-altered-babies-sentenced-3-years-jail.

Sarlin, Jon. "Anti-Vaccine Movies Disappear from Amazon after CNN Business Report." *CNN Business*, 1 March 2019. https://www.cnn.com/2019/03/01/tech/amazon-anti-vaccine-movies-schiff/index.html .

Selisker, Scott. "'Stutter-Stop Flash-Bulb Strange:' GMOs and the Aesthetics of Scale in Paolo Bacigalupi's *The Windup Girl*." *Science Fiction Studies*, vol. 42, 2015, pp. 500–518.

Willett, Walter, Johan Rockström, Brent Loken, Marco Springmann, Tim Lang; Sonja Vermeulen, Tara Garnett, David Tilman, Fabrice DeClerck, Amanda Wood, Malin Jonell, Michael Clark, Line J. Gordon, Jessica Fanzo, Corinna Hawkes, Rami Zurayk, Juan A. Rivera, Wim De Vries, Lindiwe Majele Sibanda, Ashkan Afshin, Abhishek Chaudhary, Mario Herrero, Rina Agustina, Francesco Branca, Anna Lartey, Shenggen Fan, Beatrice Crona, Elizabeth Fox, Victoria Bignet, Max Troell, Therese Lindahl, Sudhvir Singh, Sarah E. Cornell, K Srinath Reddy, Sunita Narain, Sania Nishtar, Christopher J. L. Murray. "Food in the Anthropocene: The EAT–Lancet Commission on Healthy Diets from Sustainable Food Systems." *The Lancet*, vol. 393, 02 February 2019, pp. 447–492. www.thelancet.com.

Wilson, E. O. *On Human Nature*. Harvard University Press, 1978.

———. *Half-Earth: Our Planet's Fight for Life*. Norton, 2016.

Chapter 8

The BBC Drama Series
ShakespeaRe-Told and Eric Yoshiaki
Dando's *Oink, Oink, Oink*

Iris Ralph

Withstanding the setting of Eric Yoshiaki Dando's novel about pigs, *Oink, Oink, Oink* (hereafter, *Oink*), the main events of which take place in the East Asian country of Japan (and in Australia, a country in the wider Asia Pacific region), and notwithstanding the obviously other than East Asian setting of Mark Brozel's contemporary film adaptation of *Macbeth* for the BBC series *ShakespeaRe-told*, a film inspired by the porcine species, the appearance of the two texts here, in a collection of essays devoted to East Asian ecocriticism, fits the anthology for reasons relating to the industrial farming of pigs.[1] China now is the largest industrial producer of pigs, no longer vying with and having forged ahead of the United States and Brazil in the procuring of pigs for pork.[2] The purpose of this chapter is to bring more attention to, in literary studies contexts, the porcine species and the global corporate capital livestock genetics companies that heavily invest in plying pig flesh. Pigs are prolific planetary beings, yet very little is known about them outside of the facilities where they are industrially slaughtered for processing as food and experimented on to enhance humans in transgenic research ("exchange of genetic material between different species" [Twine 54]). Billions of anthropogenic pigs are farrowed annually in those facilities. The unspoken *raison d'être* for that birthing is akin to the tectonic plates that comprise the earth's lithosphere. Floating the food and biomedical industries but concealed from everyday view, the reason is that pigs (along with most chickens on the planet as well as cows and goats) have no value beyond that of usefulness to the human species. Global corporate capital livestock genetics companies piggy-back particularly heavily on swine and stash in their piggybanks substantial profits from riding "high on the hog." The material reduction and moral debasement of pigs by those companies has intensified as pig populations have exploded throughout the world. *Oink* represents the dark, cannibalistic and eco-phagic, side of the bedfellow industries of industrial pork production and porcine genome research. Brozel's *Macbeth* raises challenging questions about mainstream attitudes of condescension toward veganism and vegetarianism. Both texts address further the denial of language to nonhuman animals. That engagement, as this chapter also will elaborate on, albeit too briefly, ties to work in the area of biosemiotics that refutes narrow, anthropocentric, understandings of and claims about language, translation, and communication.

Porcine texts from the last century include William Golding's *Lord of the Flies*, a post-World War II, ecophobic and dystopic novel about a group of schoolboys stranded on an island among its resident wild pigs in the Pacific Ocean in the aftermath of nuclear war; E. B. White's children's fiction, *Charlotte's Web,* about several farm animals who include the wise Wilbur, a livestock pig; the television series *Sesame Street*, starring the thespian Miss Piggy; Dick King-Smith's novel, *The Sheep-Pig*; director Chris Noonan's film, *Babe* (an adaptation of King-Smith's novel); Patrick McCabe's novel, *The Butcher Boy*, shortlisted for the 1992 Booker Prize; and Taiwanese author Li Ang's novella, *The Butcher's Wife*.[3] The two main porcine texts that inspire this chapter, Dando's *Oink* and Brozel's *Macbeth,* are recent, twenty-first century, additions to that list.[4] A novel in which pigs and humans change places and merge identities to the extent that they become indistinguishable species, *Oink* explicitly addresses the controversial issue of xenotransplantation and other genome research projects that involve experimenting on pigs. In Brozel's *Macbeth*, the director relies on several porcine tropes in articulating a famous argument that Jacques Derrida made about the unlikelihood of a vegetarian or vegan being a "*chef d'Etat*" (qtd. in Calarco 132). Setting the film in a city in the United Kingdom at the turn of the twentieth century, the filmmakers cast the titular figure, ("Joe") Macbeth, in the role of a head meat chef and "kitchen warrior." A virile and masculine figure whose totemic animal is a swine, Joe trains the restaurant owner's son, Malcolm Docherty, in culinary carnism.[5] Malcolm, an ex-vegetarian, is squeamish about preparing dishes of animal flesh, but to abstain from that is to forfeit the claim to his father's highly successful restaurant business. Malcolm's recidivism ensures a place for him at the tables of high society. As the film represents, abstaining from eating animal flesh is a general impediment to attaining and wielding socioeconomic power.

Oink's first person "porcine-hominoid" narrator, Squirly Fern, has grown up in Tokyo, where his mother, the daughter of a poor Okinawa pig farmer, is employed by a Tokyo fish cannery. In the late 1970s, the sixteen year old Squirly is invited by his Australian father to move to Australia. Squirly accepts the offer, moves to Melbourne, and becomes caught up in his father's projects, transgenic experiments using pigs. Transgenic research (and genome research, more broadly) has brought "into relief the porosity of previously cherished species boundaries…underlining similarity between ourselves and other species," but it has been guided from its inception virtually solely by the desire "to establish a fixed and special 'human'" relative to other species (Twine 31). The biotechnology has benefited humans in the area of research that involves pigs in particular because they share close genetic ties with humans (Baker 67; Carr 84). Already, there are proposals in place to use pigs in xenotransplantation, which involves partially humanizing pigs, or transgenically "enhancing" them, so as to prevent the rejection by the human recipients of the donor pigs' organs (Twine 53–54, 69, 80). Those proposals have raised in turn queries about human–pig ratios, or extent to which a human who has been enhanced by porcine flesh is more porcine than human. *Oink* is a dark romp down the brightly lit, labyrinthine (and increasingly publicly inaccessible) halls of transgenic research involving pigs and xenotransplantation, and a critique of the moral vacuums of that research.

Oink also is about transgenic research as that relates to the use of the term "enhancement" by livestock genetics companies. Livestock genetics companies disingenuously use it in their transhumanist posthumanist projects of anthropogenically producing tailor-made breeds of pigs (and other animal species).[6] A cheap substitution has taken place here. The "enhancement" of pigs refers to their moral and material debasement. If one is to be honest about the kinds of augmentation seen in livestock genetics companies, then one ought to use the tern enhancement in reference to the growth of those companies and the expansion in the ability to manipulate pig flesh. Other terms that livestock genetics companies use in the business of putatively improving pigs include "SuperMom" pigs and "Meidam DL GP" pigs (Twine 110, 112). The latter is used in advertising a pig breed that also is called the "mother of all mothers" (110). It is bred for its high number of teats, high number of offspring, and docility (112). Another pig breed, the "PIC410," a "hybrid boar," is advertised for "combin[ing] high primal yields, leanness and robustness with the excellent growth rate, feed conversion, carcass leanness and meat quality of the…leading terminal boar[s]…the PIC337" (110). The governments that we elect have been complacent or resistant in addressing moral questions relating to intensive genomic interferences in pigs' lives, from the artificial insemination of the animals at a young age, to the separation of piglets from their young mothers immediately after birth, to the termination of young pigs in a slaughter house or biomedical laboratory. *Oink* represents and satirizes that nonchalance and refusal in such memorable characters as Squirly Fern, Squirly's father, Pauline, and Dr. Akiyoshi.

Squirly's father works for the corporate capitalist industrial behemoth "MonsantoTM" (Dando 11). In the 1960s, Monsanto™ sends him to an "American military think-tank in Okinawa" (11). There, he meets the poor daughter of a pig farmer and, on "a dirty weekend to Mt Fuji," impregnates her (16). Soon thereafter, he abandons her, a woman whom he pejoratively calls "[t]he pig girl" (29). He returns to Australia and begins to experiment on pigs in the back garden of his home in the quintessential, middleclass, staid, and affluent Melbourne suburb of Ashburton. Since his work for Monstanto™ in Australia and the military think-tank in Japan, he has become interested in the porcine species, the "most convenient human-like test subjects for the benefit of human research" (59). His work leads him and a longtime associate, Dr. Akiyoshi, on a "gene-splicing" quest "for the most human pig" (58).

Squirly's father is especially obsessed with one pig, "P-42," whom he later renames "Pauline," a seductive dominatrix with "fluttering . . . long lashes" (61). Pauline leads other transgenic pigs in the laboratories in the back garden of Squirly's father's home to revolt against the human scientists who fiddle with their lives. Secretly "educat[ing] themselves" (61), they overpower the scientists employed by Squirly's father, "[fatten] them up in solitary metal enclosures," feed them what they have been fed, and eat some of them (62). Squirly's father negotiates with the pigs and a treaty of sorts is signed. However, it is unclear now who is a pig and who is a human, for in the course of the experiments that the humans have performed on the pigs (and those the pigs have performed on the humans) everyone involved has become "part pig" (63).

Squirly also will become "part-pig" when he undergoes a xenotransplantation operation (Dando 63). Indeed, his father already is a transgenic porcine-hominoid, having undergone "a [porcine] liver transplant in the late 1980s" (108). Scientists "grew a custom designed liver for him in one of his transgenic pigs and [he has] been using it ever since" (108). (Pauline now is "pressuring" Squirly's father to have more vital [pigs'] organs installed in his body" [108].) In the case of Squirly, Dr. Akiyoshi, the distinguished "protégé" of a scientist in Japan who worked on inventing and refining biological weapons during World War II (72), discovers and removes a malignant cyst in Squirly's thigh and "bog[s] up the hole in [Squirly's] leg with pieces of bone from a transgenic pig (107).

Oink's "MonsantoTM" is a naked allusion to the actual company of Monsanto and the questions that it and other biotechnology food companies are raising about "capitalization transgressing distinctions of human-animal and medical-agricultural" (Twine 110).[7] Squirly's father's experiments on pigs to reduce the fat content of their flesh, manipulations that result in such pigs as the uncommonly "lean" and "skinny" Pauline (Dando 35, 78, 85), are a direct allusion to the controversial drug ractopamine hydrochloride, developed by the pharmaceutical global giant company Eli Lilly and subsequently used in pork production "to promote leanness in pigs" (Twine 110). Pauline's unusually high number of teats is an allusion to livestock genetics companies' intensive selective breeding of sows so as to increase the animals' reproductive capacities (which correlate with the number and size of the animals' teats). "I have never seen so many breasts," Squirly says of his stepmother (Dando 77). The hominoid-porcine offspring that Squirly's father and Pauline produce, and their successful efforts in cloning Squirly, carry further allusions to current research in the areas of cloning and transgenic experiments on animals.

Squirly's father's ambitious plans to increase Australia's consumption of domestically produced pork (classified as a "red meat") is another marbling of the plot of *Oink* that ties to global pork production and consumption. Australians are "the largest consumers of meat in the world per capita, consuming almost twice the average of consumers in the United Kingdom," and their appetite for meat has continued to grow (Chen 41-2).[8] However, in recent decades, Australians have moved "away from red meat and towards chickens" (41). Squirly's father's plans to increase Australians' consumption of pork alludes to that actual shift as well as to Australia's modest amount of pork production relative to beef and sheepmeat production, which far outstrip pork production. In 2018, Australia was the third largest exporter of beef in the world, after Brazil and India, and the largest exporter of sheepmeat ("The Red Meat Industry"). "This nation has been riding on the sheep's back for long enough. . . . Time to ride into the new economy, ride piggy-back," Squirly's father says to his son (Dando 63). He "slaps [Squirly] really hard on the back, 'Ha! Piggy-back!')" (63). Squirly's back indeed is a "piggy-back." The xenotransplantation operation has made Squirly a transgenic porcine-hominoid animal.

Oink is a pulp fiction, science fiction, and postmodern fiction. The character of the voluble Pauline identifies the novel with another genre, that of the bestiary or beast fable, a minor but persistent literary form. Up through the end of the thirteenth century, beast fables were common throughout Western Europe. They became a marginal form of expression after that time. In England, the most celebrated beast fables that appear after the thirteenth century include Geoffrey Chaucer's "The Nun's Priest's Tale" in *The Canterbury Tales* and Robert Henryson's "The Cock and the Fox." Both texts were written for adult, highly literate and culturally elite readers; more commonly, beast fables after the thirteenth century took the form of elementary texts for teaching Latin to school children (Greenblatt 299). Writers, under the spell of Renaissance humanism, perhaps did not consider beast fables to be eminently suitable for addressing serious subject matter. Today, bestiary texts continue to identify with children's literature. Humanist beliefs about the speech of animals persist. We tend to think of speech, language, and sophisticated forms of communication as being available only to humans and so distinguish humans from other species. *Oink* is a text that constitutes the testing of that humanist grip.

Another bestiary text, *Translations from the Natural World* by Les Murray, one of Australia's most distinguished poets, implicitly questions the widespread, institutionalized denial of speech to the nonhuman world. In the poem, "Pigs," the eponymous hero recounts the last moments of his life as he is dragged along a cement floor, hung upside down, and scalded by boiling water while still half-conscious. Delivered in the "pungent dialect . . . of a hog" (Chiasson 138), the poem commits the pathetic fallacy, as bestiary texts do. The fallacy reflects a presumptive knowledge of animals on our part but also the strong hold that animals have on us even as we seem fully intent on annihilating, oppressing, and meretriciously speaking for them. Animals speak to us despite our efforts to silence them and deny them language. They inspire the discursive subjects of the poems in Murray's collection, those of Dando's novel *Oink*, and those in myriad other narratives of "interdependence" and "unlooked-for-relation . . . dispersing the human without disembodying" (Cohen 26). Such texts remind us to ask again why it is that we "suppose all things a resource for use, abuse, and transformation, and therefore blind ourselves to all matter's vibrancy" (26).

Perhaps no discipline has told us more about "all matter's vibrancy" in the context of the study of language than biosemiotics. A seminal figure in it is Wendy Wheeler, author of, among other important studies, *The Whole Creature: Complexity, Biosemiotics and the Evolution of Culture;* "The Biosemiotic Turn: Abduction, or, the Nature of Creative Reason"; "'Tongues I'll Hang on Every Tree': Biosemiotics and the Book of Nature"; and "Natural Play, Natural Metaphor, Natural Stories." In the last listed work, which appears in a collection of essays devoted to material ecocriticism, Wheeler repeats and extends a mantra for biosemiotics thinkers. It is that "human language is just [one] evolutionary part of a vast global web of semiosis encompassing all living things—from the smallest cell to the most complex multicellular organism" and "all life, not just human life and culture, is semiotic and interpretive" ("Natural Play" 71; 69). Such refrain and argument works to "put humans and human poesies and techne back in nature where they belong as evolutionary developments" (69), "erases the false sharp modern distinction between mind and body" [and "nature and nurture" and "materialism and idealism"] (69), and "obliges us radically to reconsider what we might mean when we talk about mind, consciousness, and intentionality" (69). Most importantly, such claim invites us to reconsider interpretations of "nonhuman creaturely" language that are dependent upon anthropocentric definitions of language or sign systems (69).

The study of human languages is crisscrossed with many paths, each one of which carries losses as well as gains (Flys-Junquera and Valero-Garcés 191). The study of nonhuman forms of language is as variegated as the study of human languages, and as exciting. Biosemiotics is contributing significantly to our understanding of nonhuman communication, language, and translation. Translation studies might also contribute to the as yet nebulous area of study of nonhuman languages. At the very least, translation studies holds out potentially useful lessons and tips to scholars interested in the languages of plants and animals, and in the future it may become a substrata of a larger area of study that engages with the translation of languages across species as well as languages within species other than the human species. The turn that translation studies has taken in recent decades promises that widening.

The purely linguistic basis of translation studies in the 1960s and 1970s gave way in the 1980s and 1990s to interest in the social and cultural context of translation as well as "the agency of the translator" (Bermann and Porter 3). In the wake of that turn, scholars from a range of disciplinary backgrounds—not only linguistics, but also literary theory, sociology, anthropology, cultural studies, and information technology—"enrich[ed] the field" (3). What scholars also began to emphasize after the 1960s was "different constructions of the ethics . . . of translation." That cynosure enriched and was enriched by "poststructuralist views of language," "postcolonial views of literature and culture," "gender and sexuality studies," and new frameworks of knowledge proposed by sociologists (3). For example, Jacques Derrida, Roland Barthes, and other poststructuralist thinkers argued that there is no "original" text, or that within any putative "original" text there is "ongoing intertextuality, and that readers of a text can no longer rely on or turn to a single 'author' or 'source' for a given text" (3). Edward Said, Gayatri Spivak, Homi Bhabha, and Edouard Glissant gave to the field "new, more ethically and politically inflected ways to read language, text, and translation in colonial and postcolonial contexts . . . No longer could translations be viewed as abstract linguistic entities subject to pure descriptive analysis" (4). In gender and sexuality studies, such figures as Spivak and Judith Butler opened translation studies scholars to rethinking "the way in which gender hierarchies and heteronormative assumptions affect a wide range of cultural practices, including translation and the role of the translator and reader" (4). In sociology, Pierre Bourdieu and others influenced translation studies scholars to widen their investigations beyond the translator and author of the text (4). Translation studies scholars no longer underestimate the complexities and compromises involved in the work of translating human languages. The languages of the nonhuman world continue to be viewed through narrow anthropocentric lenses. We have barely begun the monumental task of translating the languages of the nonhuman other in ways that avoid primitive comparisons between those languages and human languages.

At the same time, we can learn from the errors of scholars who study human languages and the translation of them. Just as those scholars no longer view translation as something that is purely linguistic (immaterial), unidirectional, and separate from the domains of gender, ethics, politics, and economics, so scholars and other thinkers who are interested in nonhuman forms of language and communication are recognizing that they are not merely primitive material operations. If understandings of nonhuman animal and plant language continue mostly to be saddled with reductive, "orthodox materialist" claims (Wheeler "Natural Play," 68), then one day those understandings may deepen and broaden out as a result of work in biosemiotics (and other areas inclusive of plant studies, animal studies, zoology, and ethnology). That work irrefutably reflects that nonhuman animals (and plants) possess language and the ability to communicate and translate language as that refers to complex processes of semiosis. Plants and nonhuman animals thus also have or exhibit will, altruistic behavior, "creative reasoning" (Wheeler, "The Biosemiotic Turn," 270-82), and capacity for "play" (Wheeler, "Natural Play" 75-8). While those functions take different forms and expressions than those found in humans, they are not ontologically inferior to them.

The principal moral costs of industrially producing pork and experimenting on the porcine species include denying the species any right to existence on its own terms, material, social, linguistic, and so forth. The chief material costs are global warming (Cummins; Estok, *The Ecophobia Hypothesis* 92-3). Pork products from industrial facilities one day might carry warning signs, similar to those on tobacco products, that alert consumers to the negative impact that industrial animal farming is having on the planet's lungs (forests and wetlands) and circulatory systems (waterways and seas). Already, the health industry is advising consumers to reduce their intake of red meat. (That is a salutary step forward for pigs, cows, and sheep; unfortunately, that same step is a leap backward for chickens, for consumption of chickens has escalated as that of pigs has stabilized in some countries.) In the future, the health industry may join with climate change scientists in recommending that consumers back off on their consumption of industrially farmed animal flesh so as to protect and preserve the body at the corporeal level of the planet. In major pork-producing and economically impoverished states in the United States, waterways are heavily polluted by facilities where pigs are industrially produced and slaughtered (McCorkie). In East Asia, the South China Seas now are among the most polluted waters on the planet because of the acceleration of the industrial farming of pigs (and chickens) in China and Vietnam (McCorkie). The United States gave birth to industrially farming animals in the second half of the twentieth century (McCorkie). Subsequently, despite their awareness of the huge environmental costs and the equally mounting moral questions surrounding the massive incarceration of species, major stakeholders exported the industrial methods of procuring pork to East Asia.

Mark Brozel's contemporary film adaptation of *Macbeth*, one of four films produced by the British Broadcasting Company (BBC) in a series entitled

ShakespeaRe-told, hones in on food procured from the flesh of pigs in the wider context of "the 'meatification' of global diets" in the last fifty years (Weiss qtd. in Estok, *The Ecophobia Hypothesis* 92). The film also addresses a related concern, the unlikelihood of a vegetarian or vegan wielding power in the position of a nation-state leader.[9] Vegetarianism and veganism are tolerated in many societies but not touted as part of any national or global agenda. Few who practice it are elected leaders in society. (Two notable exceptions are Jeremy Corbyn, leader of the Labour Party in the United Kingdom from 2015 to 2020; and Al Gore, the 45th Vice President of the United States.) By and large, society has accepted the "carno-phallogocentric" tradition of animal sacrifice. In *Zoographies: The Question of the Animal from Heidegger to Derrida*, animal studies scholar Matthew Calarco explains Jacques Derrida's neologism, "carno-phallogocentrism." Derrida had used it to highlight the three dimensions of "classical conceptions of subjectivity": "the *sacrificial* (carno), *masculine* (phallo), and *speaking* (logo) dimensions" (Calarco 131). That framework or "metaphysics" of subjectivity works to exclude not just animals from the status of being full subjects but other beings as well, in particular women, children, various minority groups, and other others who are taken to be lacking in one or another of the basic traits of subjectivity. Just as many animals have and continue to be excluded from basic legal protections, so, as Derrida notes, there have been "many 'subjects' among mankind who are not recognized as subjects" and who receive the same kind of violence typically directed at animals" (Calarco 131).[10] In other words, seemingly perdurable constructions of the subject, or the "multiple axes of exclusions that have functioned historically in the development of the metaphysics of subjectivity," represent that "being a carnivore is at the very heart of becoming a full subject in contemporary society" and "[p]articipating, whether directly or indirectly, in the processes and rituals of killing and eating animal flesh is . . . a necessary prerequisite of being a subject" (132). "Those individuals who, through eating a vegan or vegetarian diet" and supporting animal rights organizations, "seek to resist carnivorous practices and institutions" do not meet the dominant criteria of "being a subject" (132) "[C]arnivorous sacrifice is essential to the structure of subjectivity" (Derrida "Force of Law" 953).

In "Eating Well," another essay by Derrida that Calarco cites, Derrida states: "[t]he subject," typically male, "accepts sacrifice and eats flesh" (114). In that same essay, he rhetorically asks: "Who would stand any chance of becoming a *chef d'Etat* (a head of State), and of thereby acceding 'to the head,' by publicly, and therefore exemplarily, declaring him- or herself to be a vegetarian? The *chef* must be an eater of flesh" (114). In an early scene in Shakespeare's *Macbeth* (1.2), a captain in King Duncan's army reports on the fierce battle taking place between Duncan's Scottish forces and the invading army from Norway and the bloody execution of the Scottish traitor MacDonald by Macbeth:

For brave Macbeth—well he deserves that name—
Disdaining Fortune, with his brandished steel
Which smoked with bloody execution,
Like Valour's minion carved out his passage
Till he faced the slave—
Which ne'er shook hands nor bade farewell to him,
Till he unseamed him from the nave to th' chops,
And fixed his head upon our battlements. (1.2., lines 16–23)

In Brozel's film, in the scene that corresponds to this scene, the head meat chef, Joe, demonstrates to his staff how to dissect a pig's head, an animal stand-in for the character of MacDonald in the original play by Shakespeare. The staff include Malcolm, a former vegetarian who shows consternation during the demonstration. As the film hints at, Malcolm must overcome his distaste for animal flesh if he is to accede to the throne of his father's restaurant business. Docherty's restaurant caters to upper-class carnivorous consumers in a grimy and gritty downtown area of a late twentieth-century city. Duncan Docherty, the owner of the restaurant and the star celebrity of the business's televised cooking show, has succeeded in overcoming the handicap and stigma of a poor, rural and working class, Irish background. "Joe" (Macbeth), a Scotsman, holds the highest position in Duncan's employ. He is a master chef, a "kitchen warrior." "Billy" (Banquo), Joe's closest friend, also is a master chef. Malcolm, a trainee, is Duncan's son and the sole heir to the business. The two men whom Joe frames for the murder of Duncan are illegal migrant workers from the former Yugoslavia. They correspond to the two King's servants who are charged with regicide in Shakespeare's tragedy. Macduff, in Shakespeare's play the Scottish thane who helps Malcolm and his brother Donalbain defeat Macbeth, is the restaurant's English head waiter. The three weird sisters (or "witches") are queered working class garbage men. The battlegrounds where the Scottish and Norwegian armies battle in the opening scenes of Shakespeare play correspond in the film to the basement kitchen of Duncan Docherty's restaurant. The witches' heath is the back streets of the city where the bin men collect municipal waste, and a land-fill, where the bin men dump the municipal waste. Uniting all of the foregoing content is the central trope of the porcine species, a prominence apparently inspired by the speech of the three weird sisters in an early scene of Shakespeare's play: FIRST WITCH: "Where hast thou been sister?": SECOND WITCH: "Killing swine" (1.3).

In the scene in Brozel's *Macbeth* that corresponds to the scene in Shakespeare's play where Macbeth and his wife honor King Duncan with a banquet, Duncan holds a banquet to celebrate the news that his restaurant has been awarded the highly coveted three-star Michelin rating. He recollects two childhood memories in solemn, dignified, and sincere if also somewhat portentous speech:

> My ma woke me up twice in my childhood in the middle of the night. Once to see Neil Armstrong step out onto the surface of the moon and once to watch my father slip into the pig shed at dawn and put a knife in a sleeping Tamworth.

At this moment in the speech, it seems that for Duncan, whose career has rested on killing swine, there is little that is farcical or facetious about his implicit comparison between the momentous event of the first "landing of the moon" and the zillionth killing of a pig, "a sleeping Tamworth." However, when he perorates on the significance of the latter event, he treats it with deprecating humor. He jests to his audience that afterward he asked his father if he could "come out in long trousers," a reference to the tradition of donning long pants and abandoning shorts when a male child becomes an adult. Duncan's speech draws titters from his audience right up to his final words: "I've been out of shorts ever since."

As Duncan delivers his speech on an upper floor of the restaurant, Joe and his team dice, chop, slice, steam, boil, fry, grill, and bake mountains of flesh from the bodies of animals—namely, rabbits, fish, goats, sheep, chickens, and cows—in the basement kitchen. Duncan's speech represents how proud he is of his rural Irish roots, and of the culinary path that catapulted him out of poverty. The speech also alerts viewers to the ethical dilemmas and paradoxes surrounding animal killing. Witnessing and participating in animal killing is hardly an insignificant rite of passage from childhood to adulthood yet is reduced to levity and flippancy.

In another scene in the film, Joe appears *sans* shirt sleeves, revealing on one of his biceps a distinct tattoo of the head and prominent ears of a wild boar. After he and his wife Ella ("Lady Macbeth") murder their boss, for Joe wants to take over the restaurant before Duncan can pass the business on to Malcolm, Joe meets (or hallucinates meeting) the three bin men a second time. They assure him that unless "Pigs . . . fly!" no-one will find out that Macbeth murdered Duncan (and blackmailed a hotel worker, who served jail time in the past for committing a homicide, into murdering Banquo and the wife and children of the head waiter Peter Macduff). In the final scenes in the film, MacDuff, who by now knows Joe is the murderer of his wife and children and Banquo, confronts Joe in the basement kitchen of the restaurant after summoning the police to arrest Macbeth. The two men scuffle and Macduff mortally wounds Joe. As he hears the sounds of the wings of a police helicopter landing on the roof of the restaurant or adjacent rooftop, Joe realizes that the bin men's prophesy, "Pigs will fly," has been fulfilled.

The word "pigs" is a common pejorative for the police or "the law." As it appears in the bin men's prophesy, "Pigs will fly," it also conjures up pigs that have been genetically modified to have flying capacities. The bin men's prophesy carries an allusion, perhaps one that the filmmakers did not intend, to animal enhancement in the food and biomedical industries. It thus also brings to mind Dando's novel. In *Oink,* Squirly's father has bred "a flock of pigs with wings" (Dando 105). He does that, as other humans in actuality do that, "just to see how clever he is" (105). "A lot of money is spent on keeping the flying pigs high as kites," Squirly says (105). Uncle Barry, Squirly's father's brother, has experimented with so many drugs that he has grown a "bionic tail" (73). Flying pigs and bionic tails are the stuff of transgenic realities as well as fictions.

Readings, the Melbourne-based bookseller, advertises *Oink* as "a surreal, black comedy, a seductively hip, hilariously funny satire of popular culture and consumerism" and a "savage modern fable about science, family, television, and love gone wrong." *Oink* also is about killing swine and the heavy investment in that killing by the global corporate capital food industry and the biomedical companies that have converged with that industry. Brozel's *Macbeth*, a film that takes direct inspiration from Shakespeare's tragedy, highlights, too, the preponderance of pigs in those industries. Pigs populate East Asia as well as Australia and the British Isles in the present century. They are everywhere, outside and inside our bodies, the foods we eat, the literature we read and write, and the language we speak.

NOTES

1. See "ShakespeaRe-Told: Macbeth." Films Media Group, 2020, https://films.com/id/17014.

2. See the U.S. National Pork Board website's link, "Top 10 Pork-Producing Countries."

3. For a discussion of Ang's text in the context of pigs "fake meat" and temple worship, see Huang.

4. South Korean director Bong Joon-ho's brilliant film, *Okja*, is another recent addition.

5. The term "carnism" is from Melanie Joy's *Why We Love Dogs, Eat Pigs, and Wear Cows* (2009).

6. For a discussion of the area of posthumanism known as transhumanism, see Wolfe (xiii–xiv; xx; xv).

7. For a comprehensive account of Monsanto's controversial practices, see Robin.

8. In developing countries, annual per capita consumption of meat "has doubled since 1980," and throughout the world meat production "is projected to double by 2050" ("Meat & Meat Products"). Today, pig flesh accounts for over 40% of the "world meat intake" ("World Per Capita Pork Consumption").

9. See Weiss (4). For a discussion of vegetarian ethics and Shakespeare, see Chapter 4, "Pushing the Limits of Ecocriticism: Environment and Social Resistance in *2 Henry VI* and *2 Henry IV*" in Estok's study, *Ecocriticism and Shakespeare: Reading Ecophobia* (49–65).

10. The source of the quote is "Force of Law" (951).

WORKS CITED

Ang, Li. *The Butcher's Wife* (1983). Translated by Howard Goldblatt and Ellen Yeung. Northpoint, 1986.

Baker, Kim. "Home and Heart, Hand and Eye: Unseen Links between Pigmen and Pigs in Industrial Farming." *Why We Eat, How We Eat*, edited by Emma-Jayne Abbots and Anna Lavis, Ashgate, 2013, pp. 53–73.

Bermann, Sandra, and Catherine Porter. "Introduction." *A Companion to Translation Studies*, edited by Sandra Bermann and Catherine Porter, John Wiley & Sons, 2014, pp. 1–11.

Bong, Joon-ho, director. *Okja*. Produced by Dede Gardner, Jeremy Kleiner, et al. Distributed by *Netflix*, 2017.

Brozel, Mark, director. *Macbeth*. *ShakespeaRe-told*. Produced by BBC Ireland and the BBC Drama Department. Distributed by Acorn Media, 2005.

Calarco, Matthew. *Zoographies: The Question of the Animal from Heidegger to Derrida*. Columbia University Press, 2008.

Carr, Rachel. "100% Pure Pigs: New Zealand and the Cultivation of Pure Auckland Island Pigs for Xenotransplantation." *Animal Studies Journal*, vol. 5, no. 2, 2016, pp. 78–100, ro.uow.edu.au/asj/vol5/iss2/5.

Chiasson, Dan. "Fire Down Below: The Poetry of Les Murray." *The New Yorker*, 11 and 18 June, 2007, pp. 136–139.

Cohen, Jeffrey Jerome. "Posthuman Environs." *Environmental Humanities: Voices from the Anthropocene*, edited by Serenella Iovino and Serpil Opperman, Lexington (Rowman and Littlefield), 2016, pp. 25–44.

Cooney, Joan Ganz, and Lloyd Morrisett, creators. *Sesame Street*. Produced by Samuel Gibbon and Jon Stone, 1969–2015.

Cummins, Ronnie. "How Factory Farming Contributes to Global Warming." *Ecowatch*. ecowatch.com/how-factory-farming-contributes-to-global-warming-1881690535.amp.html.

Dando, Eric Yoshiaki. *Oink, Oink, Oink*. Hunter Publishers, 2008.

Derrida, Jacques. "'Eating Well,' Or the Calculation of the Subject." *Who Comes After the Subject?*, edited by Eduardo Cadava, Peter Conor, and Jean-Luc Nancy, Routledge, 1991, pp. 96–119.

———. "Force of Law: The "Mystical Foundation of Authority."" *Cardozo Law Review*, vol. 11, nos. 5–6, July/August 1990, pp. 919–1045.

Estok, Simon C. *The Ecophobia Hypothesis*. Routledge, 2018.

———. *Ecocriticism and Shakespeare: Reading Ecophobia*. Palgrave Macmillan, 2011.

Flys-Junquera, Carmen and Carmen Valero-Garcés. "Translation." *Keywords for Environmental Studies*, edited by Joni Adamson, William A. Gleason, and David N. Pellow, New York University Press, pp. 189–191.

Golding, William. *Lord of the Flies* (1954). Penguin Books, 2017.

Huang, Su-hsin. "The Authenticity of Fake Meats. *Interdisciplinary Studies in Literature and Environment*, vol. 19, no. 4, Autumn 2012, pp. 719–731, doi.org /10.1093/isle/iss108.

Joy, Melanie. *Why We Love Dogs, Eat Pigs, and Wear Cows*. Conari Press, 2009.

King-Smith, Dick. *The Sheep-Pig*. Gollanz, 1983.

Macbeth, ShakespeaRe-Told. Directed by Mark Brozel. Produced by BBC Northern Ireland and BBC Drama Department. Acorn Media UK, 2005.

McCabe, Patrick. *The Butcher Boy*. Pan Books, 1992.

McCorkie, Don, and Audrey Brohy, directors. *River of Waste: The Hazardous Truth About Factory Farms*. Distributed by Cinema Libre Studio, 2009.

"Meat & Meat Products." Food and Agriculture Organization of the United Nations (FAO), 2020. fao.org/ag/againfo/themes/en/meat/home.html.

Murray, Les. "Pigs." *Translations from the Natural World*. Isabella Press, 1992, p. 36.

Noonan, Chris, director. *Babe*. Produced by Bill Miller, George Miller, and Doug Mitchell. Distributed by Universal Pictures, 1995.

Readings. "Oink, Oink, Oink: A Savage Modern Fable." *Readings Pty Ltd.*, readings .com.au/products/3092082/oink-oink-oink-a-savage-modern-fable.

"The Red Meat Industry." Meat & Livestock Australia (MLA), 2018, mla.com.au/ about-mla/the-red-meat-industry/#.

Robin, Marie-Monique. *The World According to Monsanto: Pollution, Politics and Power*. Translated from the French by George Holoch. Spinifex Press, 2010.

Shakespeare, William. *Macbeth. The Norton Shakespeare*, 2nd ed., edited by Stephen Greenblatt, Walter Cohen, Jean E. Howard, and Katharine Eisaman Maus, W. W. Norton, 2009, pp. 1343–1406.

"Top 10 Pork-Producing Countries." *Pork Checkoff*. National Pork Board, 2020, pork.org/facts/stats/u-s-pork-exports/top-10-producing-countries/.

Twine, Richard. *Animals as Biotechnology. Ethics, Sustainability and Critical Animal Studies*. Earthscan, 2010.

Weiss, Tony. *The Ecological Hoofprint: The Global Burden of Industrial Livestock*. Zed Books, 2013.

Wheeler, Wendy. "The Biosemiotic Turn: Abduction, or, the Nature of Creative Reason." *Ecocritical Theory: New European Approaches*, edited by Axel Goodbody and Kate Rigby, University of Virginia Press, 2011, pp. 270–282.

———. "Natural Play, Natural Metaphor, Natural Stories." *Material Ecocriticism*, edited by Serenella Iovino and Serpil Oppermann, Indiana University Press, 2014, pp. 67–79.

———. "'Tongues I'll Hang on Every Tree': Biosemiotics and the Book of Nature." *The Cambridge Companion to Literature and the Environment*, edited by Louise Westling, Cambridge University Press, 2014, pp. 121–135.

———. *The Whole Creature: Complexity, Biosemiotics and the Evolution of Culture.* Lawrence & Wishart, 2006.

White, E. B. *Charlotte's Web.* Harper & Brothers, 1952.

Wolfe, Cary. *What is Posthumanism?* University of Minnesota Press, 2010.

"World Per Capita Pork Consumption," *Pork Checkoff.* National Pork Board, 2020, pork.org/facts/stats/u-s-pork-exports/world-per-capita-pork-consumption/.

Chapter 9

The Logic of the Glance

Nonperspectival Literary Landscape in *Wildfires* by Ooka Shohei

Kenichi Noda

THE IDEA OF LANDSCAPE IN MODERN
JAPANESE LITERATURE

In general, the term "landscape," or 風景 (*fūkei*), as it has been translated into Japanese, is the one which was adopted during the modernization process that took place in the late nineteenth century Japan, soon after the Meiji Restoration, the political revolution that took place in 1868. After that historic event, modern Japan started a process of importing a lot of words, terms, and ideas, as well as modern technologies and institutional reformations, coming chiefly from western countries. The term "landscape" was among this group of imported terminology, and it strongly influenced and activated many writers and poets aiming for creating what we currently call "modern" Japanese literature. The fact that one of the conspicuous indicators of the *modernity* of modern literature in Japan was the introduction of the idea of perspective representation in landscape depiction reveals the fact that the pseudo three-dimensional visual device called "perspective," which has supported modern visual culture, had not been dominant in the premodern literature and aesthetics of Japan. The acquisition of the perspective descriptive mode introduced a foremost novelty into Japanese literature in the late nineteenth century.[1]

Landscape is a "secondary nature" or "second nature," and it signifies a view of the natural world that has been preorganized, structured, stylized, and *textualized* as "landscape" by our senses, together with a set of aesthetics, or ideology, or tradition or culture, that has been incorporated into our gaze.[2] Contrary to the landscape representation, the natural environment *as it is*

151

should be called "prelandscape," which refers to something unarticulated by language; a shapeless nebulous state from which "landscapes are created by men intent on ordering and shaping space for their own ends." (Stilgoe, 3)[3] It was only after the concept of landscape was introduced into Japanese literature that "modern" literature became possible. Such literary transition from premodern to modern entailed a big challenge, not only regarding how to write about the landscape, but also regarding what kind of language should be used to describe it.

This "grafting" of modern onto premodern aesthetics has played an important role in modernizing Japanese literature. But even after this modernization was almost completed, the adoption of new conceptions of literary landscape did not necessarily solve the underlying contradictions between the old and the new writing styles. Generally, in the history of modern Japanese literature, the natural history essay *Musashino* (1898) written by Kunikida Doppo (1871–1908) has been considered one of the typical and earliest examples that shows how the modernization of landscape depiction was contrived.[4] Another example is found in the novel *Broken Commandment* (1906) by Shimazaki Toson (1872–1943). Both of them contributed greatly to the reformation and modernization of landscape portrayal, owing in part to their technical acquisition of the new way of looking at and writing about the landscape, deeply influenced by European writers, and they opened the way to modern Japanese literature.[5]

However, there also existed a group of writers who voiced opinions *against* the concept of landscape that came to characterize the modernity of modern literature. They disputed the very idea of landscape, in accordance with their insistence that the new idea was not so much a universal, but rather a specific one in its mode of landscape representation, and that modern Japanese literature had two different writing styles regarding the concept of landscape. In this essay, I would like to argue how these writers presented a different idea of landscape from the one held by the normally regarded "authentic, modernist" writers. Such an alternative view of "landscape" can be seen in works such as Natsume Soseki's *Sanshiro* (1912), Dazai Osamu's *Tsugaru* (1944), and Ooka Shohei's *Wildfires* (1952).[6] Among these works, particularly important is Ooka's *Wildfires,* mainly because this novel, published seven years after the end of World War II (WWII), includes a story telling how insignificant the aesthetic and perspectival view of landscape was for a Japanese soldier who stood on sentry duty in the battlefield on one of the Philippine Islands during WWII. His modernist landscape, as it were, was shattered into many pieces, like that depicted in the picture window painted by the Surrealist artist René Magritte.

POSTWAR LITERARY LANDSCAPE

This essay is primarily devoted to a discussion of Ooka Shohei's novel *Wildfires*. *Wildfires* has been regarded as one of the representative works of postwar Japanese literature. "Postwar literature" refers not only to literary works produced during the "postwar" period immediately after 1945; it is also applied to works that reflect the authors' direct war experiences and ones in which wartime experiences comprise the historical background. *Wildfires* was published in 1952, only seven years after the end of WWII. This novel won a considerable reputation and high evaluation from contemporary literary critics and general readers because it presented a so-called "battlefield realism," which had never appeared in modern Japanese literature. *Wildfires* is often said to be a semi-nonfictional novel based on the author's military experience as a soldier during WWII in the Philippines, and it gives the finest description of the reality of the battlefield, presented through the mode of empirical reportage.[7]

Since it is a story written by an ex-soldier who went through the war and battlefield experiences, this novel deeply imprints on readers a sense of the "post-warness" that existed before the formulation of the fundamental concept of postwar literature that emerged in the earlier period of the 1950s. This novel could not have been written without the end of the war; that is, without Japan's defeat in WWII. Although it is not an antiwar novel in a strict sense, it offers an intense criticism of the wartime regime and the social order of Japan. In one sense, the defeat of Japan, with its serious political and military consequences, brought forth the postwar Japanese literature which entirely reconsidered what the nation and society should have been, and what it should become. So the novel *Wildfires* emerged as a new and different type of literary work in postwar Japan. This essay's primary question is particularly directed at the following question: Where can we locate the "post-warness" and newness in this novel? If its novelty is a product of its eliminating and discarding of something old, what did this work exclude and newly acquire? Through an analysis of this masterpiece of postwar literature, this essay shows what Japanese postwar literature acquired along with this new sense of "post-warness."

First, I would like to draw attention to a cluster of subjects or themes in *Wildfires* which have previously been pointed out by many critics and scholars, so as to outline the general traits this novel predicates. *Wildfires* has been characterized as (1) a realistic war narrative, placed in a genre called the "war novel," (2) a documentary fiction, partly reflecting the author's personal experience as a soldier, (3) a story of absurdity, depicting the doubly desperate situation (of rout and sickness) of the protagonist as a sick soldier in the battlefield, (4) a story of strife within and escape from a small group of colleague soldiers who keep fleeing from invisible enemies surrounding them, (5) a story of potential cannibalism and survival in a hopeless predicament in hunger, (6) a story that questions the ethical problems concerning the involuntary murder the protagonist committed, and (7) a story of deep skepticism about God and faith, resulting in despair and madness. This list could be much longer, but the above seven points have been particularly pointed out by many critics. Although the list covers a variety of issues, ranging from considerations of its genre as a "war novel" or "documentary fiction" to ethical, religious and psychological problems, it is likely that none of them would have been fully expressed if Japan had won at the war. Japan's defeat made the genre called "postwar literature" possible, and all the items in the list are important viewpoints to be considered when discussing *Wildfires*. However, one other important thing to be discussed is not contained in this list. This is the problem of landscape description.

Although many scholars and critics have discussed this novel so far, strangely enough, not much attention has been paid to the problem of landscape and how the literary landscape is portrayed in this work. The only exception is found in the critical essay "Density of Solitude: On Ooka Shohei's *Wildfires*" (1957), written by Hino Keizo, a well-known novelist and critic. His essay urged readers to pay more attention to how the landscape was described in this story than to what the story dealt with as subjects or themes.[8] This was because the writing style that Ooka Shohei employed in his delineations of the battlefield landscape looked to Hino quite different from those in the landscapes expressed in previous modern Japanese literature. He found a uniqueness in Ooka's depictions. Hino proposed another way to read *Wildfires,* a way of reading it in terms of literary landscape, since the most fascinating aspect of this novel was, to him, its way of organizing literary landscape; a way which conflicted with that found in other modern Japanese literature. This essay hopes to build upon and extend this discussion.

WHY WAS THIS NOVEL TITLED "WILDFIRES"?

Ooka concisely epitomized the plot of the novel in his essay "The Intention of *Wildfires.*"[9] He writes that this story is about "a defeated soldier tormented by his confused senses and feelings while wandering in a tropical forest." (133) The place is in the tropical forest, where a diseased soldier is wandering. His wandering process brings him "confused senses and feelings." The loss of the sense of order, physically or psychologically, is the major hardship the soldier suffers. What, then, took place in the soldier's mind, and what kind of confusion did he experience in the forest?

The story narrates the experience of this lone, diseased soldier, wandering through the tropical forest on Mindoro Island in the Philippines. The place is a part of a small island, covered with tropical forest, at the time when the US army started to attack the island that had been occupied by the Japanese army. The soldier in this situation is described as a sort of traveler visiting a foreign country. Despite the urgency of the military rout taking place in the middle of the battlefield, the soldier remains somewhat of a visitor or traveler hanging around a foreign place that is unknown to him. Certainly he is also a soldier, and never a simple traveler. But we should not lose sight of his being a sort of "traveler" visiting a foreign country in the story. In this sense, *Wildfires* can be considered a travel narrative as well as a war narrative, because travel writing is a story of movement based on encounters with the unknown.

This Japanese soldier/traveler is surrounded by tropical landscape, which is completely unknown to him, and in it he sees "wildfires" wherever he goes.

> From the dry river bed at the bottom of the cliff rose a single pillar of black smoke. It came, no doubt, from one of the bonfires one saw so often in the Philippines at this time of the year. Since we had landed in the islands fires like this, in which the Filipinos burned the husks of their corn after harvest, seemed to have been almost constantly on our horizon. They were, in fact, our only evidence of the continued existence of the native inhabitants, who surrounded us on all sides but whom we hardly ever saw. One of the main duties of the sentries was to detect these fires and to judge from the shape of the smoke whether they were genuine bonfires for burning waste husks, or signal fires lit by guerrillas as a primitive means of communicating information to their distant comrades. (*Wildfires* 21)

Why was this novel titled "Wildfires"? The soldier had seen or witnessed "bonfires" in many places since he landed on this island. The term "wildfires" seems to suggest that, for him, there was another side (not *this* side) where enemies, like guerrillas, were hiding. To the traveler, who is one of the sentries, the foreign tropical landscape reveals a primal semiotic configuration, a dichotomy consisting of *this side* and *that side*, with an invisible borderline somewhere in between. So "wildfires" becomes a symbol standing for two different divided areas, confronting each other with hostility.

The soldier asks himself whether the fire is natural or man-caused. The identification is impossible for him since the dichotomy of *this side* and *that side* does not afford him any opportunity to cross over this strict and absolute border. "Wildfires" suggests something that keeps indicating the existence of such an unidentifiable borderline and the conceptual predicament he has fallen into. "Wildfires" continue to function as a cause of anxiety for the soldier throughout the story because it presents to him a riddle that "no visible facts can resolve" (30). The choice of the word "wildfires" reflects the author's original intention. When he selects the term "wildfires," instead of merely "fires," he presents a rhetorical hint; as if suggesting that these are natural fires and therefore accompanied by an unidentifiable, suspended sense of anxiety. Soldiers, especially defeated soldiers in battlefields, have to read every sign of the enemy's existence in order to survive. In the setting of this story, "wildfires" are the primary signs that need to be detected in order to assure their survival.

Therefore, the soldier reads the signs that a fire represents, in order to interpret whether it is natural or not. If it is *not* natural, the further question arises: Is it a simple daily fire (such as wild burning) or a fire that the "enemy" has made with some intention? The defeated soldier cannot decode the signified (*signifié*) that the fire denotes, for he cannot go over the border between *this side* and *that side*. This fact indicates that the soldier's reading never leaves the level of speculation and that it is blocked by strict limits. The soldier's every effort to read and interpret what the fires actually mean is useless.

We have observed that the lone soldier is obsessed with an anxiety caused by his inability to decode the signified meaning of the "wildfires." But we should mention another matter here; that of the nature–culture relationship, as distinguished from that of the hostile relationship which we have discussed so far. "The unfamiliar circuitous route in the forest" (15) that the soldier takes is described as follows in the story:

> It was dark within the forest. On either side of the narrow path towered huge oaklike trees; the space between them was densely filled with unfamiliar brushwood and shrubs that stretched out the tentacles of their vines and creepers in all directions. The ground was thick with moldering leaves, which here in the tropics kept falling regardless of the season; the surface of the path felt soft as rubber under my feet. The newly fallen leaves made a rustling sound that reminded me of the road through Musashino Plain at home. With my head bowed I walked along. (15–16)

The anxiety caused by the wildfires, as indefinite signs, is closely related to the place where he is. This passage explicitly shows that a significant comparison between the known and the unknown was made by the soldier when he was watching the surrounding landscape. The passage portrays a process of finding similarities and differences by contrasting the known and unknown. Phrases like "huge oaklike trees," "unfamiliar brushwood and shrubs," and "moldering leaves, which here in the tropics kept falling regardless of the season," are highly suggestive of an eye for comparison. And the most important thing is that the soldier/viewer's reference point in making his comparison is the "Musashino Plain," which refers to a suburban area in Tokyo.[10]

Thus, *Wildfires* evokes its landscape through reference to a similarity/difference relationship regarding an image of the "Musashino Plain" in Tokyo. When one encounters an unknown world, he or she keeps observing the surrounding objects by means of analogies and comparisons with things already known. "Musashino" is the place that provides the soldier/viewer a frame of reference when making a comparison between the known and unknown.

VISION AND VISUALITY

The memory of Musashino intervenes in the views which the soldier actually sees in the forest of the Philippines. This kind of intervention, or doubleness, can be explained by referring to the theory about the difference between "vision" and "visuality" commonly employed by such art historians as Hal Foster and Norman Bryson. Hal Foster simply points out that "vision suggests sight as physical operation and visuality sight as a social fact" (ix).[11] And, in the same vein, Bryson adds to this the following:

> When I look, what I see is not simply light but intelligible form: the rays of light are caught in a *rets*, a network of meanings, . . . Vision is socialized, and thereafter deviation from this social construction of visual reality can be measured and named, variously, as hallucination, misrecognition, or "visual disturbance." Between the subject and the world is inserted the entire sum of discourses which make up visuality, that cultural construct, and make visuality different from vision, the notion of unmediated visual experience. Between retina and world is inserted a screen of signs, a screen consisting of all multiple discourses on vision built into the social arena. (91–92)[12]

As Bryson argues, when we look at something, what we see is not only light but also socialized "intelligible form," which holds within itself "a network of meanings," "a screen of signs," or "a screen consisting of all multiple discourses." Musashino, which is recalled by the Japanese soldier in the forest of the Philippines, functions as a site where "visuality" exists as a "cultural construct" because "the entire sum of discourses" is inserted "[b]etween the subject and the world." The landscape of Musashino, which is an aggregation of multiple socialized signs and discourses already known to the soldier, generates a frictional, conflicting movement between vision and visuality through an act of comparison.

In the dark forest, the soldier meets with the moment when visuality or the "screen of signs" is annihilated by the tropical landscape that has developed before his eyes. In this sense, the landscape that he sees in the forest in the Philippines is actually an unconsciously constructed vision of the "unmediated visual experience" that gave rise to it. And the soldier perceives this landscape as a vision that has presented itself as the unknown and unknowable otherness because it is beyond the visuality of his "cultural construct" called Musashino. This notion of a landscape as a vision perceived by "unmediated visual experience" also suggests that the dark forest evinces the alterity of landscape, which is consistent with the representation of wildfires. Thus, visuality, as a cultural and social construct, and corresponding to a known landscape, such as that of Musashino, is instantly inserted, as well as collapsed in this scene.

SENTRY'S HABIT

Ooka Shohei talked about his own sense of landscape as being derived from a "sentry's habit" in an essay he wrote about *Wildfires*.[13] The idea of the "sentry's habit" emphasizes the way soldiers in general look at their surrounding environment. For example, he writes in *Wildfires*: "[O]ne of the main duties of the sentries was to detect these fires." (21) This relates, within this story, not only to a professional "habit" that all soldiers should follow, but also to the extreme fear and anxiety they must always face. This sentry's habit affords a special kind of sense toward landscape in the soldier's behavior.

> A successful infantryman must look at nature only from the standpoint of necessity. A gentle hollow in the ground is nothing but a shelter from artillery fire, the beautiful green fields simply dangerous terrain that must be crossed at the double. (19)

The "sentry's habit" requires the soldier to constantly look at nature out of "necessity." Aesthetics is not necessary. Detecting and avoiding any kind of danger is, first and foremost, important to any soldier. By using the terms "sentry's habit" Ooka was referring to the rejection of aesthetics in looking at landscape. We have already remarked that Mindoro Island in the Philippines is a foreign and unknown place to the Japanese soldier and that its tropical landscape is unknown, too. That is why he tried to associate the unknown landscape with the known landscape of Musashino in order to overcome the sheer strangeness and otherness he felt.

Still, the surrounding tropical landscape remained a landscape as vision with no "screen of signs," with no signs of Musashino. Owing to his "sentry's habit," the landscape reflected in the soldier's sight always looked a little different from the ordinary landscape. In other words, Ooka Shohei's landscape description exhibited its own conspicuous characteristics. The passage that follows is one of such typical examples:

When I left the village, I could see the fields spread out in all directions. About half a mile straight ahead to the north they were bounded by a forest. On the right, the flat paddy fields stretched into the distance as far as the volcanic mountain range that formed the spine of Leyte Island. One low branch of the mountains extended to the left and formed a backdrop to the forest ahead of me; it was curved gently like the smooth back of recumbent woman. Far to the left its undulations came to an end with a small hummock, and at this point I could make out a rapid river about twenty yards wide. The mountains rose steeply again on the opposite bank and followed the river downstream. Beyond these mountains must have lain the ocean. (13–14)

The features of this landscape description are shown in a line of sight, which constantly moves from front view to right view and then from right to front and from front to left. The front view expressed by the words "straight ahead" and "ahead of me" in the passage designates an arbitrary visual front from which the sentry looks. Being completely devoid of a sense of directions like East, West, North, and South, the viewer's objectivity seems to become poorer and weaker, and the subjectivity becomes strengthened. But nothing can give validity even to this seeming subjectivity because a sentry's watchfulness is essentially passive. And, paradoxically, it is this passivity that maximizes his ability of information gathering as well. Yet, this arbitrariness may point to the source of anxiety in the sentry's mind as well.

The views of front, right, and left are combined into a certain sequential movement in this landscape description. Thus, the landscapes in *Wildfires* are represented as two-dimensional, planar ones, lacking three-dimensional perspectival depth. How are we to define this mode of landscape discourse? Hino Keizo sharply pointed out in his essay in 1957 that "this novel's description of nature lacks the mode of perspective, and, instead, there is left–right symmetry." While visual perspective is an idea that emphasizes the depth and directions of gaze, the left–right symmetric description in *Wildfires* exclusively displays "the configuration of objects on the flat surface with no depth" (Hino 98).

Let us compare the following landscape in Kunikida Doppo's *Musashino* with the landscape in *Wildfires*:

Now go back and take the other fork. Immediately the trees thin out and a broad field stretches before you. You walk down a gentle slope and everywhere there is miscanthus, its tips glittering under the sun.

Beyond the miscanthus, the fields, and beyond them a clump of low trees. Fleecy clouds gathered on the horizon and perhaps a chain of mountains, deceptively taking on the colour of clouds themselves. (p. 106)[14]

We can easily observe that this landscape is depicted in a linear perspectival structure. A single scenery running from the viewer's eye in the foreground extends linearly toward the farthest part in the background near the "horizon" where "clouds," and "a chain of mountains" gather. The scene gazed on by the viewer composes, specifically by the use of the word "beyond" reiterated twice, a perspectival depth that converges in a vanishing point or "horizon" in the background.[15] The perspectival method employed in this work provides a sharp contrast with the left–right symmetry of the landscape description in *Wildfires*.

Perspective is one of the most evident elements we can use to clarify the *modernity* of modern Japanese literature. Kunikida's *Musashino* was published in 1898 and Ooka's *Wildfires* in 1952. About a half century passed between the writing of these two works. *Musashino* has been highly regarded as one of the representative innovative literary works in the history of modern Japanese literature, mainly due to its landscape depiction based upon linear perspective. The "sentry's habit" expressed in *Wildfires* motivated the author to partly change the landscape description from linear perspective mode to the left–right symmetric mode after fifty years had passed and after the end of WWII.

TIME IN LANDSCAPE

The left–right symmetry method is one of the most conspicuous traits in *Wildfires*'s literary landscape. In addition to the passage quoted (above, p. 159), the following two passages illustrate further examples of how this methodology is preferred in this novel.

> At the entrance to the forest the road divided into two. Straight ahead was the path that crossed the hills and led directly to the opposite valley where lay the hospital. The left-hand fork circled the hummock by the river and eventually emerged into the same valley. The path over the hills was, of course, far shorter, but having already taken it three times in the past twenty-four hours, I decided on the spur of the moment to try the unfamiliar circuitous route. (15)

> The forest stretched to my left in the direction of the sea and disappeared in the distance. Ahead of me the plain rose and fell like a huge sand dune until half a mile away it reached a bare rock formation, which blocked my view like a screen. Halfway between me and the rock the grass was burning across the plain in a strip some sixty yards wide. There was no one in sight. (28)

The adverb phrases like "[S]traight ahead," "[T]he left hand fork," "to my left," "[A]head of me," and "[H]alfway between me and the rock," create a visual movement as if corresponding to the shifting directions of the sentry's eyes. The frequent switching from one view to another seems to create a certain dynamism in the landscape description. The frequent use of the shifting directions suggests that the entire landscape in these paragraphs involves the passage of time. It means that every sight the sentry's eyes capture belongs to a different spot of time. Every turn of the eyes produces a different view or snapshot, only to reveal the lapse of time. In other words, these landscapes described in the left–right symmetry method comprise an aggregation of multiple, different spots of time, corresponding to each instant's viewpoint. The French philosopher Maurice Merleau-Ponty might well say about what happens here: "When our gaze travels over what lies before us, at every moment we are forced to adopt a certain point of view and these successive snapshots of any given area of the landscape cannot be superimposed one upon the other." (40)

On the contrary, the linear perspective mode lacks a sense of time. The landscape captured by the linear perspective is somehow timeless or motionless in time (Jay, quoted below 7) because it does not involve aggregated multiple points of view. In his discussion of the "Cartesian Perspectivalism" (Foster, xi), the art historian Martin Jay writes succinctly about what matters most to the "scopic regime" that has become established and dominant in the modern visual world.

> Significantly, the eye was singular, rather than the two eyes of normal binocular vision. It was conceived in the manner of a lone eye looking through a peephole at the scene in front of it. Such an eye was, moreover, understood to be static, unblinking, and fixated, rather than dynamic, moving with what later scientists would call "saccadic" jumps from one focal point to another. In Norman Bryson's terms, it followed the logic of the Gaze rather than the Glance, thus producing a visual take that was eternalized, reduced to one "point of view," and disembodied. (Jay 7)

According to Martin Jay, the modern habit of "Cartesian Perspectivalism" presupposes that the perspectival eye is singular, and essentially static, rather than dynamic, and that it is unblinking and fixated. An immobile single eye dominates and gives shape to what it looks at, which is the landscape resulting from "Cartesian Perspectivalism." Similarly, Gina Crandell, a landscape architect, writes, "[L]inear perspective creates the illusion of seeing spatial relationship from a single vantage point." (Crandell 8.) In this context, when what we see becomes "landscape," it should more accurately be called an illusion created by "[T]he three-dimensional, rationalized space of perspectival vision." (Jay 6)

We can see another significant idea in the latter part of the passage above. That is Martin Jay's reference to the terminology used by Norman Bryson relating to the logic of the Gaze and the logic of the Glance.[16] Bryson discusses the linear perspective mode as being dominated by the logic of the

Gaze, and the nonperspective mode as dominated by the logic of the Glance. This distinction between Gaze and Glance is worthy of argument in regard to considering the nature of the landscape in *Wildfires*. While the Gaze in linear perspective mode only captures a momentary view, with no further extension of time, the Glance embraces a multiple time consisting of a "series of visual impressions," or "free visual impressions" (Merleau-Ponty 40–41). These "free visual impressions" are accumulated as shots taken by incessantly moving and watchful eyes like those of the sentry. Significantly enough, the author Ooka Shohei used the logic of the Glance, not the logic of the Gaze, to enable him to follow the movement of the sentry's eye and, by so doing, he could succeed in giving form to the temporalized landscape in which a part of this novel's realism is materialized.[17]

In his essay "Exploring the World of Perception: Space," Maurice Merleau-Ponty addresses the issue of how the perspectival landscape distances us from unmediated visual perception. According to him, "[T]he world of perception," or "the world which is revealed to us by our senses and in everyday life," (31) and our effort "to rediscover the world as we apprehend it in lived experience," (39) has become abstracted by the "analytical vision" or the "analytical overview" (41) which geometrical perspective induces. Furthermore, Merleau-Ponty argues that what we view as landscape represents, essentially, "a common denominator to all these perceptions" (40) by setting up "an absolute observer" (41) who is dominantly situated at the picture's center. This observer/viewer extracts "a single, unchanging, landscape" (41) from the "series of visual impressions" (41) for he/she interrupts "the normal process of seeing" and subjugates all details to "analytical vision" (40). As a result, he/she "fashions on the canvas a representation of the landscape which does not correspond to any of the free visual impressions" (41). As Gina Crandell writes, this is the time when "the illusion of seeing spatial relationship from a single vantage point" appears (Crandell 8).

TEMPORALIZED LANDSCAPE

In light of the Merleau-Ponty's argument regarding the world of perception, Ooka's logic of the Glance used in *Wildfires* is deeply related to his notion of the "birth of the landscape." Merleau-Ponty emphasized his point about such changes of perception as follows: "If many painters since Cézanne have refused to follow the law of geometrical perspective, this is because they have sought to recapture and reproduce before our very eyes the birth of the landscape." (41)

Ooka claims that when modernist landscape painting comes to be inactive, that "the birth of the landscape" is brought about. Thus, we can say that

Ooka Shohei attended "the birth of the landscape" in his effort to describe what passed before his eyes in the logic of the Glance as the sentry in the novel states: "I glanced swiftly about the plain" (94). The logic of the Glance continually operates in directing what he sees in creating a sort of temporalized landscape. Merleau-Ponty mentions Paul Cézanne because we can find, for example, in his masterpiece "Still Life with Commode" (1887–1888), an ideal model to express "the birth of the landscape," or the logic of the Glance. He writes about this sort of painting:

> Thus different areas of their paintings are seen from different points of view. The lazy viewer will see 'errors of perspective' here, while those who look closely will get the feel of a world in which no two objects are seen simultaneously, a world in which regions of space are separated by the time it takes to move our gaze from one to the other, a world in which being is not given but rather emerges. (41)

As if corresponding to this statement, Ooka writes of an often-seen landscape in a passage of *Wildfires*:

> The entire central mountain range now spread out ahead of me under the glowing morning sky—the mountain range that at this very moment was being looked at from thousands of different angles by the thousands of soldiers on Leyte Island as they faced their imminent deaths. From here I stood, its silhouette was like the humps of a camel's back. (59)

The mountain range being looked at "from thousands of different angles by the thousands of soldiers on Leyte Island" does not simply mean that a number of defeated Japanese soldiers' eyes saw it at the same moment. Rather, it means that there is, as Merleau-Ponty says, "a world in which being (is not given but rather) emerges" ahead of many and different glances. Moreover, this landscape discloses the complete loss of objectivity in soldiers' minds. The impossibility of crossing borders interrupts the soldiers' possibilities of obtaining objectivity. The soldier looking at the silhouette of the mountain range associates it with a camel's back. This fanciful simile, slightly echoing the well-known scene in Shakespeare's *Hamlet* (Act III, Scene 2), implies a world of diverse interpretations with no ultimate objective truth.[18]

The verbal landscapes which *Wildfires* creates reflect the moving eyes of the sentry, employing the logic of the Glance. This use of temporalized landscape shows explicitly how the novel acquired a sense of "post-warness" and newness that no other modernist Japanese literature had displayed, based on an alternative means of delineating landscape that differed from the modern perspectival one. Hino Keizo has suggested that Ooka Shohei's non-perspectival description is rooted in East Asian traditional culture: a particular form of landscape described in Chinese poetry and mountain–water pictures (or 山水画, *sansuiga*). Similarly, Norman Bryson has found a theoretical basis for his ideas regarding the logic of the Glance in the art of Chinese and Japanese calligraphy and landscape painting.[19] Thus, the literary landscape in *Wildfires* exhibits its newness by employing premodern literary forms.

The issue of antiperspectivalism, which is distinctly and deliberately employed in *Wildfires,* has a deep connection with the so-called "postcolonial awareness" that the author eventually acquired through and after his war experience, cruel, painful, and horrified. Those anti- or nonperspectival scenes viewed or described in this novel by an ex-Japanese soldier/author, and those sights captured by *not* his single "gaze" but his multiple "glances," reveal the author's intention of denial to the "scopic regime" which had been dominant in the *modern* Japanese literature. To put it differently, this denial shows a clear expression of Ooka Shohei's postcolonial gesture or criticism toward the Japanese colonialism which sent the soldier/author to the battlefield in the Philippines during WWII.[20]

NOTES

1. Regarding the problems of modernity and the perspective mode in modern Japanese literature, this essay profited particularly from Karatani Kojin's *Origins of Modern Japanese Literature*, specifically chapters 1, "The Discovery of Landscape" and 6, "On the Power to Construct." Also important is Fujimori Kiyoshi's discussion in his *Modernity in Narrative*.

2. The basic meaning of the word "landscape" as it is used in this essay is found in the following two definitions: "A view or prospect of natural inland scenery, such as can be taken in at a glance from one point of view; a piece of country scenery" (*Oxford English Dictionary*) and "An expanse of scenery that can be seen in a single view" (*American Heritage Dictionary*).

3. John R. Stilgoe, *Common Landscape of America, 1580–1845*, New Haven and London: Yale University Press, 1982.

4. In this essay, personal names of Japanese writers are given in the Japanese order: surname first, followed by the given name.

5. Among the influential European writers whom Japanese writers often mentioned during the late nineteenth century are Ivan S. Turgenev, William Wordsworth, R. W. Emerson, and John Ruskin.

6. The standard translated version of the novel has a slightly different title from the original Japanese one. The English version was given the title "Fires on the Plain." But in this essay I prefer to use "Wildfires" as its title because the *literal* translation of the Japanese title, *nobi*, should be "wildfires." In Japanese, the first part of this word, *no-*, can have several meanings, one of which means "wild," as an adjective, while

another meaning is "field" or "plains," and the latter part, *-bi* means "fire." This essay refers to the English version titled *Fires on the Plain* (translated by Ivan Morris).

7. Ooka is known for his nonfictional works such as *Taken Captive: A Japanese POW's Story* (1949) and *A Record of the Battle of Leyte* (1971), written extensively about his own war experience. *Wildfires* might be exceptional among his many works in the sense that, in it, he takes a fictional approach to his experience. While nonfictional works are designed to objectify his experience and memory, a fictional work like *Wildfires* subjectifies them, as this essay concludes in the final part.

8. Hino's insightful essay "Density of Solitude: On Ooka Shohei's *Wildfires*" was first published in 1957, only five years after the publication of *Wildfires*.

9. "The Intention of *Wildfires*," in *The Collected Works of Ooka Shohei*, Vol. 14.

10. Musashino Plain is the setting for the natural history essay titled *Musashino*, written by Kunikida Doppo. Musashino is the suburban area that extends through the western part of present-day Tokyo. Kunikida's *Musashino* was written and has been enthusiastically read as a modern literary text by so many people that this area has come to be regarded as a site with a mytho-poetical aura. Ooka Shohei was born and brought up in this area, so his memory about this place shares many points with that of Kunikida Doppo.

11. Hal Foster, "Preface," in Hal Foster, ed. *Vision and Visuality*.

12. Norman Bryson, "The Gaze in the Expanded Field," in Hal Foster, ed. *Vision and Visuality*.

13. "On Sentry's Eyes" in *A Collection of Ooka Shohei's War Narratives: A Story of Shoes*, pp. 96–97.

14. Kunikida Doppo, "Musashino."

15. Karatani Kojin, *Origins of Modern Japanese Literature*, especially chapter 6, "On the Power to Construct," extends the discussion on the significance of perspectival "depth" in relation to the linear perspective in the modern literary history of Japan.

16. For more detailed theoretical information about the distinction between the Gaze and the Glance, see Norman Bryson's *Vision and Painting: The Logic of the Gaze*.

17. Martin Jay's *Downcast Eyes* briefly examines how war or battlefield experience affects "the visual experience and the discursive reflection on that experience." (212).

18. Ooka wrote a novelized version of *Hamlet*, titled *The Journal Written by Hamlet*, published in 1980. See Cyrus Hoy, ed, *William Shakespeare: Hamlet* (Second Edition): Act III, Scene ii, p. 57. Hamlet's reference to a camel is found in the following: HAMLET Do you see yonder cloud that's almost in shape of a camel?/ POLONIUS By th' mass, and 'tis like a camel indeed./ HAMLET Methinks it is like a weasel. /POLONIUS It is backed like a weasel./ HAMLET Or like a whale?/ POLONIUS Very like a whale.

19. See Norman Bryson, chapter 5, "The Gaze and the Glance" in *Vision and Painting: The Logic of the Gaze*.

20. This work was supported by JSPS KAKENHI grant number 15H03201.

WORKS CITED

(All the information on books and articles written in Japanese was translated into English by the author. And all the translation from books and articles written in Japanese were done by the author.)

Bryson, Norman. *Vision and Painting: The Logic of the Gaze*. New Haven: Yale University Press, 1983.

Crandell, Gina. *Nature Pictorialized: "The View" in Landscape History*. Baltimore and London: The Johns Hopkins University Press, 1993.

Foster, Hal, ed. *Vision and Visuality*. Seattle: Bay Press, 1988.

Fujimori, Kiyoshi. *Modernity in Narrative* (in Japanese). Tokyo: Yuseido Publishing Company, 1996.

Hino, Keizo. "Density of Solitude: Ooka Shohei's *Wildfires*." (in Japanese). In *Collection of Critical Essays on Ooka Shohei* (Edited by Kamei Hideo), Wildfires: *Critical Approaches to Modern Literature 16*. (in Japanese) Tokyo: Kuresu Publishing Company, 2003.

Hoy, Cyrus. *William Shakespeare: Hamlet* (Second Edition). New York and London: W. W. Norton & Company, 1992.

Jay, Martin. *Downcast Eyes: The Denigration of Vision in Twentieth-Century French Thought*. Berkeley; Los Angeles; London: University of California Press, 1993.

Karatani, Kojin. *Origins of Modern Japanese Literature* (Translated by Brett De Bary). Durham; London: Duke University Press, 1993.

Kunikida, Doppo. "Musashino." In *River Mist and Other Stories* (Translated by David G. Chibbett) Tenterden, Kent: Paul Norbury Publications Limited, 1982.

Merleau-Ponty, Maurice. *The World of Perception* (Translated by Oliver Davis). London and New York: Routledge, 2004.

———. "Cézanne's Doubt." In *Sense and Non-sense* (Translated by Hubert L. Dreyfus and Patricia A. Dreyfus). Illinois, 1964.

Noda, Kenichi. "Battlefield Experience and Musashino in Ooka Shohei's *Wildfires*." (in Japanese) *Tokyojin* No. 410, Tokyo: Toshi Shuppan, 2019.

Ooka, Shohei. *Fires on the Plain* (Translated by Ivan Morris). Tokyo: Tuttle Publishing, 2001.

———. "The Intention of *Wildfires*." (in Japanese). In *The Collected Works of Ooka Shohei*, Vol. 14. Tokyo: Chikuma Shobo, 1996.

Scolari, Massimo. *Oblique Drawing: A History of Anti-Perspective*. Cambridge, Massachusetts: The MIT Press, 2012.

Stilgoe, John R. *Common Landscape of America, 1580–1845*. New Haven; London: Yale University Press, 1982.

Chapter 10

The Paradox of Aerial Documentaries

Eco-Gaze and National Vision

Sijia Yao

The seemingly progressive heyday of capitalism could be the apocalypse of all human beings given the "ecological disequilibrium" (Guattari 1989), "hyperobjects" (Morton 2013), and the "worldwide ecological crisis" (Žižek 2010). An aerial viewpoint can actively reverse or fasten such a progressive deterioration. In this chapter, I will examine the first Taiwanese aerial documentary, *Beyond Beauty: Taiwan from Above* 看见台湾 (Chi Po-Lin 2013), in comparison with two similar documentaries in mainland China, *China from Above* 鸟瞰中国 (Kenny Png, Klaus Toft 2015) and *Aerial China* 航拍中国 (Yu Le 2017), to analyze the cinematic aesthetics and the consequently ethical tension between eco-consciousness and nationalism. I further problematize and complicate this aerial view to call for a new way in which micropolitical action of art can be applied to macroplanetary ends of healing. The elusive logic inherent in aerial view ultimately leads us to raise the urgent question: How can we as cultural producers and consumers resolve the tension between nation-based modern development and earth-bound humanity that knows no boundaries, be it national, ideological, social, or otherwise?

Taking the same bird's-eye point of view as well as employing the same advanced aerial-filming equipment, the Chinese directors, however, produce aesthetically and ethically different cultural products. The difference demonstrates the different stages of social developments in the two regions. Taiwan, approximately twenty years earlier than mainland China in launching its economic globalization must decades earlier recognize the devastating pollution and energetic depletion as the necessary evil in the economic progression. More importantly, the comparison of the three aerial documentaries will facilitate us to find various, nuanced aesthetic ways in which aerial exploration of landscapes can be appropriated to meet different social ends. It is the paradoxical power of a high vantage point of view in either drone surveillance, national governance, or planetary awareness that renders this project significant and highly relevant to our time.[1]

THE POWER OF AERIAL VIEW

All the three films are strongly featured with an aerial viewpoint regardless of their different aesthetic modes and ethical focuses. The direct translation for the title of *Beyond Beauty* 看见台湾 is "to see Taiwan," while the other two Chinese film titles literally mean "bird's-eye viewing China," and "aerial-filming China." The aerial view, salient and significant, has drawn the critical interests of scholars, among whom Roland Barthes is perhaps the first. Barthes begins his demystification of the Eiffel Tower with the bird's-eye view from the top of the Tower.

> The bird's-eye view [. . .] gives us the world to read and not only to perceive; this is why it corresponds to a new sensibility of vision; in the past, to travel [. . .] was to be thrust into the midst of sensation; the bird's-eye view, on the contrary, permits us to transcend sensation and to see things in their structure (Barthes 9).[2]

For Barthes, the aerial view offers "a new perception" that can create aesthetic distance to allow the observer to read and decipher the place, in this context, the city of Paris (9). The observer's point of view is privileged because it involves a process of "decipherment," the "activity of the mind, conveyed by the tourist's modest glance" (10). The urban geographer David Harvey indicates that, from the panoramic and God-like viewpoint, we can possess the place "in imagination instead of being possessed by it" (Harvey 1).[3] For Barthes and Harvey, the aerial view gives rise to a new intellectual mode of interpreting the world and simultaneously can easily invoke intensive feelings of awe, hubris, or simply visceral excitement.

On the other hand, in today's era of technically assisted governance and drone wars, the aerial viewpoint is naturally encoded with national self-portrait, surveillance, and military hegemony. The most extremely negative example of this view is the remote killing depicted by Rey Chow:

> War can no longer be fought without the skills of playing video games. In the aerial bombings of Iraq the world was divided into an above and a below in accordance with the privilege of access to the virtual world. Up above in the sky, war was a matter of maneuvers across the video screen by US soldiers who had been accustomed as teenagers to playing video games at home; down below, war remained tied to the body, to manual labor, to the random disasters falling from the heavens (Chow 35).[4]

The technologically innovative aerial view, in this sense, carries the violent and policing nature of dominance and manipulation. It echoes with Paul Virilio's vigilant statement, "totalitarianism is latent in technology" (104, qtd. Kellner).[5] Hence, this point of view from the above cannot escape the exploitation of the ruling class or dominant discourse.[6]

The gaze from above over the landscape, therefore, encapsulates a paradox: triggering an affective connection to nature, but simultaneously inviting a transcendental impulse of colonizing or dominance. Denis Cosgrove already points out the opposing dichotomy of aerial visions in his book, *Geography and Vision*: the aerial images can impart "a new sensitivity to the bonds that bind humanity to the natural world" but at the same time encourage "an Apollonian perspective" of "rational control" and "planning" (Cosgrove 89).[7] Viewing from above, therefore, is a question of ethical gesture. The aerial view itself cannot be ethically evaluated without knowing the ways in which it is used and applied by filmmakers.

In visual studies, the way of looking/seeing resembles the way of reading/interpreting in literary studies; thereby, it is inevitably an ethical act or a criticism. Given the power and fluidity of the vantage point, the choices made by the aerial documentary director directly determine the degree of honesty and justice in the revelation of the place, the nation, and the world in general. As early as in 1938, Martin Heidegger already declared the advent of the "Age of the World Picture," when "the world" is "conceived and grasped as a picture [. . .] set up by man, who represents and set forth" (129).[8] An ethical revelation and construction of an in-depth world picture demand a critical eye and ethical stance. Moreover, documentary films always serve the purposes of public education, political mobilization, and civic engagement. The aerial documentary filmmaker, therefore, is empowered by the aerial vision to conceive and represent the world picture. A director's cinematic interpretation of the world through aesthetic practices and ethical propensities directly determines the messages and affects received by spectators. Hence, it is urgent to study the aesthetic and ethical languages in recent Chinese aerial documentaries.

ECO-GAZE, FILIAL SORROW, AND *BEYOND BEAUTY*

As the first Chinese aerial documentary, *Beyond Beauty* (2013), directed by the environmental activist Chi Po-Lin and produced by the world-class Taiwanese director Hou Hsiao-hsien, strategically employs an aerial viewpoint to interpret the picture of Taiwan in a highly ecocritical vein. This film is opposed to the usual appropriation and exploitation of the aerial view in documentaries that encompass utilitarian anthropogenic needs, such as tourist pleasures, a progressive social agenda, geographical education, or national cohesion as exemplified in the other two documentaries to be discussed later. Reading Taiwan in a Lacanian fashion, Chi Po-Lin's camera, charged with ecocritical energy, gazes upon the invisible from above and emphasizes the visible that people usually choose to evade or escape from.

The cinematic narrative is primarily structured with a contrast between sublime nature and its scarred body afflicted by the inhabitants. Its strong sense of place and dialogic interaction reverses and challenges the conventional narrative that privileges human dominance over nature in the Anthropocene epoch. Chi creates an eco-documentary that subordinates humans to nature and landscapes, calling Taiwanese people "temporary guests of this land (Chi 1: 19)," and "children of the island," which is indicated in the closing sequence. The body of the mother(land) as well as the filial tension with her children, as the core narrative, generates a deep affect of filial sorrow. Chi's visual rhetoric is not only an addition to the environmentalists' tradition, but more importantly offers an alternative to breaking the paradoxical restriction of this tradition through promoting intergeneration, cross-species, and intercorporeality.

It is this filial affect that drives Chi to predicate an ecocritical revelation and interpretation of Taiwan on its status as an island in a planetary sense, rather than a regime or social venue in an anthropogenic sense. It is also this affect that persuades and moves audiences to reconsider their interrelation and visceral bond to nonhuman entities and take further actions of healing and renovating. Borrowing the rich and complex cultural meanings a high vantage point of view connotes and its close relation to filial connection in traditional Chinese literature, Chi embodies cultural specificities in his documentary to invigorate and complicate the aerial view with a Chinese heritage. Chi also offers a strong visual statement that is bolstered by a traditional Chinese logic: *ru qing ru li* 入情入理, which means to persuade with feelings and reasons. Feeling precedes reason in a Chinese literary tradition, thereby reversing the conventional emphasis of cognitive reason in argument.

Ancient Chinese instilled the feeling–reason logic in to their cultural definition of an aerial view. In Chinese tradition, the term, *denggao*登高, literally means to climb to a high vantage point, for instance, the top of a hill/mountain or the top of a pagoda. This cultural term usually connotes a far-away viewpoint or a deep sorrow caused by a longing for the beloved. Chi inherited the default cultural connection between the sorrowful feeling and the high vantage point to enrich the aerial view in his documentary narrative.

It is easy to extrapolate the argument by quickly tracing back this specific tradition in classical Chinese literature. In Denggao (ascending), one of his poems composed in *Kuizhou* 夔州 (today's Fengjie) in 767 A.D., Du Fu, one of the best Chinese poets in the Tang dynasty, expresses his deep sorrow after climbing high and viewing the desolated autumn landscapes on the Yangtze River. He laments, "ten thousand li away from home, I am grief-stricken by the autumn scenery and my fate of drifting. Old and ill, I climb to the top alone." (万里悲秋常作客，百年多病独登台).[9] At the age of fifty-six, the unfortunate poet underwent the sudden fall of the Tang dynasty (following the "An Lushan Rebellion," 755–763 CE) and the ensuing chaos, poverty, and illness. It is the heartbreaking pain and the penetrating grief that crown this poem as the best septa-syllabic (or seven-syllable) regulated verse in China. It is the high vantage point of view that invoked a profound sense of sadness in Du Fu as well as us as his readers. Also, this high vantage point of view is intimately related to a deep longing for family and the beloved. Du Fu followed the tradition to climb high on Chong Yang Festival 重阳节 or the Double Ninth Festival (the ninth day of the ninth month of the lunar year) to miss his family and long for his home in a long distance. Hence, the high vantage point of view in the Chinese tradition usually delivers a deep sorrow or melancholy for parents, lovers, and family members.

Furthermore, climbing the heights of pagodas or mountains also involves bodily exertion, absorbing varying degrees of the elements, and an ineluctable human sense of humility amidst natural grandeur. It brings forth the humans' reverence for humility vis-à-vis the nonhuman world to fulfill the process of transcorporeality in which the human, the place, and the nonhuman species cross the boundaries to find each other and to form a whole through visceral experiences. For Chi, the sorrowful, humble filial gaze from above is cast upon the homeland that is allegorized as a suffering mother. The human and the nonhuman, therefore, are transcorporeal through a profound filial affect.

THE STRUCTURE OF ECO-NARRATIVE IN CHI'S AERIAL DOCUMENTARY

Juxtaposing a magnificently virgin nature with a severely bruised body of a motherland,[10] this auteur film incarnates its filial gaze in "a bird" or "a cloud," flying high above to explore this "land of beauty and sadness," as the voice-over introduces in the beginning of the film. Because its aerial gaze is part of nature, the documentary liberates itself from a conventionally anthropocentric narrative to elevate nature to an equal and transcorporeal relation with the human. A most recent example of a disappointing aerial viewpoint is Ai Weiwei's documentary *Human Flow* (2017). In this humanist, self-righteous, and well-acknowledged auteur documentary, Ai's aerial viewpoint taken by drones "at a ninety-degree angle, and from a great height (Ai 2)" dehumanizes lines of refugees "from far" as "a swarm of ants (6)," and subjugates the filming object and its significant humanist discourse to the director's own artistic and activist ego, as the French philosopher Georges Didi-Huberman astutely points out.[11] Ai's filming from high above reveals a colonizing and condescending gesture toward the human suffering no matter how sincerely and empathetically he feels for the refugees. Taking an opposing gesture, Chi lowers the aerial angle and shortens the distance from the filmed object, to establish a face-to-face, egalitarian connection to nature and landscapes. The metaphor of land as the motherly body establishes the human's visceral attachment to the earth as absolute bottom-line, which beckons an earth-bounded nature of humanity.

The first shot of the earth followed by the zoom-in on the Taiwan Island sets the fundamental note for the documentary: the planetary rather than the global. There are diverse ways of aerial viewing in *Beyond Beauty.* The primary two modes of aerial looking are the explorative, appreciative bird's-eye view over the grand nature, and the following sorrowful filial gaze at the ruined body of the motherland. Both modes of aerial viewing, taken by a bird or a descendant of this island, tightly bind humans to nature and land-scapes. The documentary begins with a slow and quiet panoramic extolling of Taiwan's mountains, farmland, ocean, and natural beauties and wonders. Although from high above, the Cineflex lens usually takes the angle less than ninety degrees to exhibit the marvelously grand structure of a river, moun-tains, and ocean in order to arouse a sense of awe and intimate connection. The patient and attentive gaze upon the body of this homeland defamiliarizes the place of Taiwan by humbly approaching tops of mountains at an eye level and attentively flying by trees, birds, deer, and creek as close as possible. Driven by an intensive curiosity and modesty, Chi's lens is shortening, even merging virgin nature and human beings. It is the way in which Chi's camera positions itself in relation to the nonhuman that distinguishes the aerial eco-gaze in this documentary from other models of aerial views.

Such aerial long-takes of the landscapes last five minutes and forty-seven seconds until Wu Nien-jen's humble and peaceful voice-over invites the audi-ence to perceive and understand their motherland in a revolutionarily new vision. "Please don't be surprised. This is our home, Taiwan. If you haven't seen it this way, perhaps that's because you didn't stand high enough." The privileged view of the almighty natural landscape accompanied by the spiritual music "Genesis" in the opening sequence underscores the primacy of origin and home. The twenty-five-minute landscape aesthetics undercut human dominion and present a magnificent body of a motherland in the first section of the documentary.

If the aerial view magnifies the power of nature's magnificence and sacredness in the first section, it also renders the environmental damages and wounds tremendously visually appalling and affectively overwhelming in the second section. From this contrast there rises up a penetrating, embodied feeling experienced by audiences or inhabitants of this island to ponder upon the relation between nature and themselves. From above, the camera with a filial gaze is witnessing ongoing crimes and caressing the mother's body. Her unfilial children, who are obsessed with profit and progress, inflicted deep scars and ugly wounds upon the motherland.

The deforestation for high-mountain tea paddles and economic plants skins off the mountain valleys and tops. The landslides caused by road building and housing expose the bones of the mountains. The countless pipes withdrawing underground water for the profitable fish industry greedily drain the blood of the Island. The toxic pollutions dye every vein of the water system. The destruction of the coastland pulls out the teeth of the land to replace them with concrete wave-breakers and artificial harbors. The wetlands, named as "kidney of the earth" (53:00) by the voice-over are clogged and consumed by shocking amounts of mudslides and garbage. The gravel mining in the mountains even cut open the bellies of the mountains to take out muscles and organs for economic development. The aerial gaze shows us a scarred body that reminds us of the pristine nature and nurturing home in the first section. The strong affect embedded in these disturbing images directly offers us visceral experiences of pain and violence. Each of these images is an aerial gaze of nature that deserves respect and care. The voice-over comments after the endless images of the "slow violence."[12]

This island has been nurturing the inhabitants with her flesh and blood generation after generation. Just like a mother who has borne too many children, her body is gradually exhausted and in great agony. However, she remains silent without any complaints, supplying her children with anything they want. Only when her children greedily squeeze the last drop of her milk, can we hear her moan gently in pain. (1:11) The bodily violence to the homeland is as unethical as the unfilial misconduct to the mother according to Chinese values and conventions. The song, "Kanjian" (看见, to see) as an integral part of the audial narrative, fully illustrates the filial regret for the damage and the intention to heal the place. "Can you see? I cannot see [. . .] everything we had is all gone [. . .] Take so much; return so little. If there is tomorrow, can we heal the wounds to let the earth be reborn?" Intertwined with the lyrical reflection are thousands of inhabitants happily living their lives on this land. The audio and the visual form a dialectical mode of statement, reminding people that their happiness is founded on the land's nurturing and motherhood. People are gazing at the camera cheerfully but also being gazed at from above. The camera directly confronts everyone who is responsible for the ungrateful deeds to the motherland with the endeavor of healing awaiting in the future.

Admittedly, the rhetoric of anthropomorphism can be paradoxical in its function. It can be a colonizing and exploitative strategy of dominance.

On the other hand, it can equally be a powerful strategy to dissect human-centered ethical dominance and reverse the human/nature dichotomy by way of undermining human dominance. In Chi's case, he blends the notion of filial piety with the human corporeal ties to nature via the trope of the maternal body. The word, *xiao* 孝 (filial piety), was inscribed on a bronze vessel that can be dated to the very last years of the Shang dynasty, over three thousand years ago. In *Xiaojing* 孝经 (The Classics of Filial Piety), it states, "夫孝，天之经也，地之义也，人之行也" (Filial piety is the law of the Sky, the righteousness of the Earth, and the conduct of the Human). Filial piety, as the root of all virtues and conduct, has been crucial to Chinese civilization. Filial piety, as the first and foremost among the hundred virtues in Confucius Chinese doctrine, has been residing in the core of the social order.[13] Moreover, the blood ties that affectively define filiality even have a more profound impact than socially constructed virtue. Chi's aerial view offers the audience bodily, highly visceral experiences to see the wounded body of the motherland from a grander scale, evoking sadness and regret.

This filial affect makes Rosi Bradotti's transversal subject feasible in practice. As Bradotti theorizes transcorporeality, she writes, "we need to visualize the subject as a transversal entity encompassing the human, our genetic neighbors the animals and the earth as a whole" (Bradotti 82). Filiality, therefore, embodies a denial of the human as a transcendent, humanist individual who is aloof above the world he surveys. In other words, filiality opens a crack for us to intimately entangle, engage, and interact with other human generations, nonhuman bodies, and the planet, and eventually abandon human exceptionalism.

The marriage of the sorrowfully filial connection and environmental justice serves as a convincing rhetorical strategy to extend the present, which is conditioned by postmodern fragments, to a holistic and continuous historical time. The backward filial gratitude to parents is extended to a forward-looking cultivation of posterity. The past is organically linked to the future with a filial blood bond. Actually, the fact that the Chinese language has no time tense demonstrates the Chinese thinking tendency: all the times are connected and overlapping. The eyes are always casting backward and forward simultaneously. This way of looking is particularly intensified by the aerial viewpoint that can temporalize the space. For Barthes,

> to perceive Paris from above is infallibly to imagine history [. . .] from the top of the Tower the mind finds itself dreaming of the mutation of the landscape which it has before its eyes' through the astonishment of space, it plunges into the mystery of time [. . .] it is duration itself which becomes panoramic. (Barthes 11)

Likewise, Chi temporalizes the place through a continuous duration of filiality and posterity. This temporal continuity and extension is reinforced by the structural contrasts, the sublime origin versus the ugly human appropriation at present; the intact beauty in the past versus the mutated injured body now and in the future. The aerial views attract meanings and contemplations to restore temporal depth, filial care, and generational continuity.

In the end, the camera follows a small group of children climbing to the top of Yu Mountain, Taiwan's highest mountain, to envision a brighter future of healing and restoration. The choir of children standing on the peak of Yu Mountain cheerfully expresses the hope for posterities' rejuvenating connection to the earth. Rather than voicing their minority ethnicity, the indigenous children singing in their own language promote a primitive time when humans believed themselves as part of the ecosystem and worshiped nature as their life givers. Inheriting ecological memories, the primitive and indigenous culture reminds us of our collective identity: children of the earth. This pristine origin extends toward a promising future. The aerial view in the beginning is composed of slow-motion close shots of the group of children. Later, the camera gradually distances itself and pans above them to celebrate the fusion of the children and the grandeur of the landscapes. The notion of posterity, is prevalently adopted as a popular environmentalist strategy, no matter how problematic and complex it is, as Adeline Johns-Putra successfully unpacks this notion in her article "Borrowing the World." Rather than "a convenient signifier," as Johns-Putra observes, posterity strategy could negatively serves, the posterity in Chi's aerial view is situated as part of the biosphere, as in/from the body of the motherland (John-Putra 6). In the last aerial shot, the relation of the children to the mountain, the sky, and the non-human entities in general is not separate, but connected and transcorporeal, as tight as family blood. Like filiality that is charged with rich affective appeal, posterity delivered at the end of the film projects a wish for ecological interdependence of the human and the nonhuman in a broader planetary sense.

NATIONAL VISION IN *CHINA FROM ABOVE* AND *AERIAL CHINA*

If the aerial gaze, filial and reflective, in *Beyond Beauty* ecologizes Taiwan, the national vision constructed by tourist pleasure, progressive reason, and state strengthening in *China from Above* and *Aerial China* elevates China to be a strong player in the game of global progress. This mode of aerial view, not charged with profound affect, is driven by man-made pleasure-seeking desire and national pride. The national vision embodied in these aerial views separates the human from others by boundaries (national, spatial, ecological),

promotes eco-technological advancements by placing price tags on nonhuman species and places, and subjects the surveyed objects to anthropogenic
needs. This mode of aerial view necessitates an effective circulation of sociopolitical agendas and promotes human exceptionalism to render environmentalist actions peripheral and optional.

Different from Chi's auteur film, both *China from Above* and *Aerial
China* were sponsored by the state and influenced by global aerial programming. The logic and narrative of the two documentaries also vastly differ
from the independent film, *Beyond Beauty*. The power of aerial view to
incite and mobilize people can serve a very different sociopolitical agenda
and rhetoric. Co-produced with the National Geographic Channel in 2013
and released in 2015, *China from Above* is the first aerial documentary in
mainland China. Following some aerial programming in Europe and the
United States, mainland China also wanted to adopt the aerial narrative to
establish a glorious image of the nation. With a strong mark of National
Geographic Channel, this documentary features accurate geographic knowledge, exciting tourist experiences, and local people's stories within various
defined places.

Similarly, another documentary series, *Aerial China*, whose trailer was pre-
miered on New York's Times Square for three days in a row in 2016, exhibits
one Chinese place in each episode to meet tourists' desire for exciting, enter-
taining experiences. Highly aware of the foreign gaze, both documentaries
are eager to occupy the center of the global stage by presenting an emerging
nation in its full splendor. The gaze from above becomes a colonizing look,
either fetishizing the land as otherness, offering a strong dose of oriental or
exotic fantasy, or declaring the unrivaled eco-tech strength of today's China.
And yet, there is a dangerous omission of nature and ecology in this official
rhetoric of the "world's factory." The exotic or socially advanced representation of China on the global stage utterly glosses over a broader understanding
of planetary coexistence.

China from Above, as an "epic journey from the air" or "aerial journey,"
takes a tourist viewpoint to expose a transformed new nation for the world
to see. This journey will allow the audience to "discover how ancient traditions, engineering, agriculture, and natural wonders shaped this great nation
and continue to forge modern-day China after centuries of seclusion," as
the voice-over introduces the film. Composed of two parts, the documentary starts with pleasurable images of the Water Festival, dazzling Shaoling
Kong Fu, giant historical projects, such as Leshan Buddha, the Great Wall,
and the ancient hanging temple. In the aerial view, the breathtaking natural
wonders, the precipitous mountains that inspired the filmmaking of *Avatar*,
and the mysterious Heaven's Gate Cave effectively quench audiences' thirst
for miracles and mysteries. The village grandma zip lining over a grand river,

the largest ice and snow sculpture site, and the boundless water rice terrace all entertain and educate audiences with extraordinary sights and information. Empowered by the aerial view, those images of indigenous traditions and wonders are ready to be sold and consumed.

The tourist gaze in the second part gradually transforms in to an admiring look at a series of hyperobjects: mega cities, the largest internal migration, the fastest bullet train, the biggest indoor swimming pool, the massive filming complex, the giant ultra-voltage cable network on precipitous mountain ridges, and endless solar parks. Everything within the aerial scope is grand-scale engineering. The gaze from above feeds on human hubris as well as encourages an illusion: humans possess omnipotent power like the Creator. This gaze has no intention to participate in Timothy Morton's prospect for an "ethics that can handle hyperobjects" (Morton 123). The Sky Eye, a spectacularly giant telescope hidden in mountains, perceived from above becomes the sublime that is created by men and taken with pride by fellow citizens. These glorified images of modern progress pronounce China's confidence and ambition in its eco-tech development now and in the future, but dissolve the bond with other human generations and other nonhuman entities.

The national vision is further deepened and circulated in a larger aerial documentary series, *Aerial China*, produced by CCTV (China Central Television), China's chief state television broadcaster, with the first season screened in 2017. Imitating the American predecessor, *Aerial America* (68 episodes in total, released in 2010), each episode of *Aerial China* is an aerial video tour of a Chinese administrative division. This huge filming project attempts to geopolitically configure and reveal a China that boasts twenty-three provinces, five autonomous regions, four municipalities, and two special administrative regions. Continuing the cartographic intention and mapping compulsion, the aerial gaze conveys a totalizing vision of a nation with clear boundaries to its neighbors.

The episode on Hainan Island begins with introducing its geopolitical significance as a national boundary marker in the South Pacific Ocean. The detailed cinematic exploration of the marginal small islands that guard the national line reminds us of the documentary's function as a nation-building apparatus. The historical sites, local animals, and indigenous culture affirm enduring and potent psychic links between the landscape and shared collective identity. Inheriting the progressive gesture and optimistic tone in *China from Above*, *Aerial China* foregrounds a sense of national unification and belonging. The aerial gaze, accompanied with an informative and appreciative voice-over, celebrates such images as a fast-racing bullet train, a large ship that holds a train segment, and an advanced rocket site. The eco-tech progress and development boost up national pride and patriotic emotions.

Delving into the same topographic site of an island as in *Beyond Beauty*, the camera gazes upon Hainan not as an integral part of nature but as a national boundary to the rest of the world. The national image of a competitive player in the international arena heavily relies on a series of aerial images, such as advanced transportation, abundant agriculture, lively cities, and hyperobjects that serve modern needs and desires.

Here, I am not framing a binary opposite in aerial expressions. Nor am I privileging one area over another by comparison. It would be naïve and ignorant if I overlook the dilemma or oversimplify the complex relation between modern development and natural environment in both mainland China and Taiwan. There are still speeches and deeds that degrade the environment in Taiwan; while in mainland China there is an increasing awareness about ecological crisis. However, my critical scope here concentrates on aesthetic strategies of the aerial view and its ethical gesture to seek the way in which aerial representations on a small scale can exert impact upon society and environment on a large scale. In the first three aerial Chinese documentaries, I find it significant and useful to decipher the paradoxical power that makes salient the intractable dilemma between nation-building and an ecological agenda.

CONCLUSION

"Every exploration is an appropriation," as Roland Barthes insightfully observes (14). If every attempt of exploration made by media will end up with appropriation, the filmmaker who takes an aerial view can easily appropriate the object to make strong arguments. Virilio proposes an analogy between directors and dictators. "For a whole generation of cinematic miracle-workers, the process of direction, even if improvised, literally took the form of revelation—that is divine action which makes known to men truths that they would not be able to discover by themselves" (Virilio 52).[14] Accordingly, a director can conceive and sell strong images of a successful nation that advocates a rational and progressive agenda but overlooks environmentalist ethics. Or, a director can choose to step back and lead us to see an outlook based on the ecological intercorporeality of human and nonhuman. Chi Po-Lin's unfortunate death in an aerial-filming accident in 2017, a mythically coincidental practice of Barthes's "the death of the author," ironically reinforces his authorial intension of ecocriticism and activism artistically and ethically.

With the coming of the space age, the world picture in the aerial view could become a paradoxical icon depending on the authorial intension and narrative structure. Aerial life in media fully charged with messages and affect could either enable or restrict the imagined national vision. The concept of planetary humanities can be interpreted and implemented through the lens of aerial view, a significant standpoint that can invite us to see, know, and meditate on holistic and ecological healings in the future. Aerial view, therefore, has already emerged as not merely a technological innovation, but as a cultural nexus where every one of us has to make a choice when we produce or receive aerial images in media. It also leads us to the inevitable dilemma, riddled with intention, between national development and planetary healing.

NOTES

1. It is indeed valuable to explore how this difference occurs despite the fact that the two entities evolved along the same Confucian ethical and cultural heritage. And yet, given the breadth and depth of this topic, I will not develop this line of exploration in this essay.

2. Roland Barthes, "Eiffel Tower," *The Eiffel Tower and Other Mythologies*, trans. Richard Howard (New York: Hill & Wang, 1982), 9.

3. David Harvey, *The Urban Experience* (Baltimore: Johns Hopkins University Press, 1989), 1.

4. Rey Chow, "The Age of the World Target: Atomic Bombs, Alterity, Area Studies," *The Age of the World Target: Self-referentiality in War, Theory and Comparative Work* (Durham, NC: Duke University Press, 2006), 25–43.

5. Douglas Kellner, "Virilio, War and Technology: Some Critical Reflections" *Theory, Culture and Society*, 16.5–6 (1999): 103–125.

6. I am not denying the fact that aerial view can serve public goods. For instance, it is the space aerial-view that enabled scientists to be among the first to discover global warming and rising sea levels. Here, I want to emphasize the dubious vantage viewpoint in its dominion power versus its revelatory power.

7. Denis Cosgrove, *Geography and Vision: Seeing, Imagining and Representing the World* (London: I. B. Tauris, 2008), 89.

8. Martin Heidegger, *The Question Concerning Technology*, trans. William Lovitt (New York, NY: Colophon, 1977), 129.

9. All the English translations for the Chinese texts in this article are mine.

10. I deliberately insert a dash between mother and land in this essay to distinguish my term of "mother-land" from the politically loaded word, motherland, with an intention to lead our attention to a broader sense of planetary existence.

11. Accessed at https://www.eurozine.com/high-vantage-point/

12. Rob Nixon, *Slow Violence and the Environmentalism of the Poor* (Cambridge, MA: Harvard University Press, 2011).

13. Wang Yongbin 王永彬 (1792–1869) in Qing dynasty writes, "Filial piety is the first and foremost among the hundred virtues" (*baishan xiao wei xian* 百善孝为先) in *Weilu yehua* 围炉夜话 (Fireside Chat). https://www.gushiwen.org/wen_914.aspx

14. Paul Virilio, *War and Cinema: The Logistics of Perception* (London and New York: Verso, 1989).

Sijia Yao

WORKS CITED

Braidotti, Rosi. *The Posthuman*. Cambridge: Polity Press, 2013.

Johns-Putra, Adeline. "Borrowing the World: Climate Change Fiction and the Problem of Posterity." http://epubs.surrey.ac.uk/814160/1/Borrowing%20the%20World.pdf

Morton, Timothy. *Hyperobjects: Philosophy and Ecology after the End of the World*. Minneapolis: University of Minnesota Press, 2013.

Virilio, Paul. *War and Cinema: The Logistics of Perception*. London and New York: Verso, 1989.

Chapter 11

Humans, Mermaids, Dolphins

Endangerment, Eco-empathy, Multispecies Coexistence in Stephen Chow's *The Mermaid*

Kiu-wai Chu

Juxtaposing real-life environmental issues with cinematic metaphors in Chinese cinema, this chapter focuses on the ecological, multispecies fable between a human and a mermaid in Hong Kong comedian Stephen Chow Sing Chi 周星驰's *The Mermaid* 美人鱼 (2016), and discusses how mythical creatures and environments in the film serves both as a critique of environmental destructions in Hong Kong caused by excessive urban developments over the past two decades; and the allegories of postcolonial Hong Kong under drastic transformations. It illustrates how traditional Chinese folk legends are reimagined and recontextualized in the film, placing cultural imaginations of (post)colonial Hong Kong within ecocritical contexts, in order to facilitate our ecocritical awareness toward planetary healing from multispecies perspectives.

The Mermaid broke numerous box office records both domestically and internationally. Released in February 2016, it received the biggest opening day and opening week of all time in China. To date, it remains the seventh highest grossing film in China's film history, receiving a domestic total gross of over USD$3.2 million, and a worldwide total gross of over USD $553 million.[1] As a rare case for comedies, *The Mermaid* was nominated for numerous awards in major Hong Kong film awards, and won the Best Director award of 2017 Hong Kong Film Critics Society. The Society rated the film highly for its environmentalist message and its critical reflection on the negative consequences of hyper-consumerism.[2] As Stephen Chow states in an interview, what he aims to convey in the movie is the message that "With human activities leading to the ever-accelerating rate of environmental destructions,

181

we should shift beyond our human perspectives, and try to see what human beings have done to the Earth, from the perspectives of other [nonhuman] species."[3]

However, by focusing on its speciesist representations, this chapter will examine why *The Mermaid*, as a commercial film with multiple environmental messages, falls short in various aspects to cultivate better viewers' ecocritical awareness and convey more effective messages in promoting planetary healing. Drawing from recent discussions on eco-cinema studies, multispecies ethnography, and Hong Kong's enviro-political situation; from the legendary half-human, half-fish creatures of Lo Ting, the endangered pink dolphins in *Sha Lo Wan*, to the comical mermaids in Stephen Chow's film, this chapter explores the extent in which mainstream fictional Chinese films could bring to light issues of multispecies connectivity, coexistence and harmony, in a region under drastic socioeconomic and environmental changes.

THE MERMAID AND ITS MULTIPLE LAYERS OF ENVIRONMENTAL MESSAGES

Stephen Chow's *The Mermaid* is a romantic comedy that tells the story of Liu Xuan, first introduced to us as a multibillionaire, a property developer characterized by arrogance and greed, and his romance with a beautiful mermaid Shan. In defiance of all rules of society, Liu spends tenths of billions of dollars to purchase the Green Gulf (*Qing Lo Wan*), a bay area with land along its shore for a large-scale reclamation project. He hires a team of scientists who brutally chase out the dolphins, fish, and all living creatures in the sea with destructive and life-threatening sonar devices, in order to facilitate their reclamation plan. We soon realize a whole community of mermaids have discreetly resided in the Green Gulf for over hundreds of years, whose existence is not known to the human beings, but they are increasingly marginalized and suffering from the consequences of water and environmental contaminations. To avoid humans' acts of pollution, environmental damages, and mermaid slaughtering, they have been living their sheltered lives in a dilapidated old sunken ship. Liu's project made their home increasingly uninhabitable. The young and beautiful mermaid Shan is sent to infiltrate the human world, disguise herself as a human in order to kill Liu. Under the deception of a narcissistic and heartless businessman, Liu is an insecure, lonely but kind-hearted person. He meets Shan on several chanced occasions and soon falls in love with the innocent mermaid-in-human-disguise, and gradually accepts her true identity of a mermaid. Liu begins to realize the serious damages his reclamation project has done to the Green Gulf, as well as to Shan and the mermaid tribe, inhabitants of the wrecked ship by the gulf. He decides to call off the reclamation work, but the idea is objected by his business partners. And, in order to continue with the project, they increase their efforts in catching and slaughtering the mermaids. To prove his love for Shan, Liu turns against his business partners, and fights bravely on the side of the mermaids and successfully saves the whole mermaid community. In the end, Liu and Shan live happily ever after.

Being a box office success in both local and international markets, *The Mermaid* tactfully delivers its ecocritical messages in two particular ways. To appeal to the broader global audience, it conveys a universally accepted environmentalist message that is based on ethical treatment of others and conservation of the environment. At the same time, the film makes implicit metaphoric reference to traditional folklores and the endangerment of pink dolphins in Hong Kong since the early 2000s, to appeal to the culturally specific Hong Kong and Chinese-speaking viewers. Through deterritorialization of narrative setting, and re-territorialization in a metaphorical sense, *The Mermaid* invites viewers to discover its multiple environmental messages that successfully address both global and local concerns, and promotes both universal environmentalism and bioregional awareness.

DETERRITORIALIZING THE
ENDANGERED ENVIRONMENT

To fully understand the significance of the deterritorialized setting of *The Mermaid*, one would have to examine the sociopolitical background of Hong Kong and its relationship with China over the past few decades. With the transfer of sovereignty from Great Britain to the People's Republic of China in July 1997, Hong Kong's time as a British colony came to an end. And yet, despite twenty plus years after its handover, and the increasing political instabilities since 2019, Hong Kong's constitutional principle of "One Country, Two Systems" and its status as a Special Administrative Region continues to distinguish it from other Mainland Chinese cities.

Since the beginning of the twenty-first century, the opening up of China's film market facilitated the rapid development of cross-regional film productions between Hong Kong and Mainland China. Joint venture productions with mainland Chinese partners have become a new trend in Hong Kong cinema, as is clearly reflected in the post-2000s works of Stephen Chow. While Chow has retained his nonsensical and comedic film aesthetics seen in *Shaolin Soccer* (2001), *Kung Fu Hustle* (2004), *CJ7* (2008), *Journey to the West* (2013), and *The Mermaid* (2016), his stories have gradually shifted from distinct Hong Kong urbanscape to increasingly fictionalized settings.[4] *The Mermaid* obfuscates the geographical boundaries and cultural differences between Hong Kong and the two Mainland Chinese cities in the South, Shenzhen and Dongguan, where the actual scenes were filmed, evident in the local vehicle number plates and mainland police uniforms that are seen throughout the film. At the same time, the sights of Hong Kong-styled tea cafes (cha chaan teng), cooked food kiosks and children playgrounds with signboards written in traditional Chinese characters that are unlikely to be seen outside of Hong Kong, are simultaneously observed. While international audience not familiar with the region would not easily identify such culturally hybrid and unfamiliar spaces that mesh together distinctive features of Hong Kong and Mainland Chinese cities, such design, which aligns to the government's grandiose plans in building the Great Bay Area by unifying Guangdong, Hong Kong, and Macau, reflects Chow's commercial consideration in envisioning a fictional urban environment of deterritorialization that blurs any distinctive difference between Mainland Chinese cities in the South and Hong Kong the former British colony, in order to ensure the film could appeal to both the Cantonese-speaking Hong Kong as well as the much broader Mandarin-speaking Chinese markets. It also results in a commercial film that appears to convey a more universalized environmental message that is generally accepted, namely the significance of environmental preservation, and the ethics of care for the nonhuman animal species.

Despite its effect in creating a deterritorialized diegetic world that goes beyond local attachment and expands people's environmental awareness toward the planetary, in the following section, we will discuss how *The Mermaid* adopts two forms of allegorical reterritorializations in order to convey more subtle but specific environmental messages to the viewers: first, with the ecological reference to local ancient folklores, and second, to offer an implicit social critique on the infrastructural developments in Hong Kong that affect the Chinese White dolphins' wellbeing.

THE LEGEND OF LO TING: MULTISPECIES IMAGINATION THROUGH ANCIENT FOLKLORES

From a small nineteenth-century fishing village to an entrepot of the British colonial era, changes in Hong Kong's history and culture were intimately associated with water. The ancestors of Hong Kong boat people came from many different ethnic subgroups who lived on the water. One of these largest groups is the Tankas (蜑家). As Hong Kong art historian and curator Oscar Ho points out,

> Ever since their arrival in Hong Kong, the Tankas have been an oppressed group. For a long time, they were not allowed to move ashore and live on land.

Even during colonial days, the British administration imposed many restrictions on the pretext that the Tankas exhibited debauched behaviour. The diaspora of the Tankas' pitiful existence, without a base that they could call home, seems an apt reflection of the fate of Hong Kong people. (Ho 172)

Since ancient time, there have been countless legends about these indigenous Hong Kong fishermen. In his studies of ancient history of Hong Kong, Ho conducted literary search on a number of allusion texts, to examine the origin of the Tanka tribe. The legend has it that the early Tankas were half-human, half-fish amphibian creatures called Lo Ting (盧亭) who lived a nomadic life in the waters near today's Lantau Island in Hong Kong, as depicted in a good number of Chinese historical documentations.[5]

According to New Sayings on Guangdong (廣東新語), the 17th century publication by scholar Qu Dajun, "Lo Ting were usually found in areas between Tai Yu Shan and Wanshan Archipelago. They were hermaphrodites with an appearance resembling humans; their hair was a burnt yellow and short, eyes were yellow also, skin a dark, yellowish complexion, and donning a tail of over an inch. Afraid of humans, they always dove into the water on seeing people. Sometimes they would float and be carried along by the waves and people would chase them because they looked so weird." (Ho 173)

As Ho suggests, these half-fish creatures were alienated by fellow villagers, and their descendants tend to hide their identity in order to avoid discrimination and isolation. It is said that many native islanders in today's Lantau Island, particularly those "who have the surname Lo, with dark skin, slightly yellow eyes, and an especially long coccyx, are offspring of Lo Ting." (Ho 176)

At the 2010 Annual Convention of American Anthropology Association (AAA), an art exhibition titled the Multispecies Salon featured works that are created based on postanthropocentric, multispecies imaginations. The exhibition also illustrates the emerging field of multispecies ethnography in western scholarship, which focuses on exploring how "the human has been formed and transformed amid encounters with multiple species of plants, animals, fungi, and microbes", and reconstruct "the understanding of 'human' communities as in reality conglomerates of human and nonhuman species that shape each other." (Heise 2016: 195) Rather than simply celebrate multispecies mingling, multispecies ethnographers are more eager to explore the question, "Who benefits, cui bono, when species meet?" (Kirksey, Shuetze, and Heimreich 2014: 2) To answer this question, scholars' multispecies studies are "collaborating with artists and biological scientists to illuminate how diverse organisms are entangled in political, economic and cultural systems." (ibid.)

Over a decade prior to the multispecies ethnography emerged in the western academic and art worlds, in 1997, Oscar Ho already curated an art exhibition titled "Hong Kong Reincarnated: New Lo Ting Archeological Find (香港三世書之再世書盧亭考古新發現)," which also illustrates multispecies imaginations in Chinese context. The exhibition displayed local Hong Kong artists' works that fabricate historical documents and archaeological artefacts to prove Lo Ting's existence, together with a series of sculptures and paintings in the theme of Lo Ting.[6] (Figure 11.1) By creating an alternative history museum of a fabricated past, Ho and the artists reimagined and represented the complex and hybridized cultural identity of Hong Kong people, at a time when Hong Kong experienced its transition from a British colonial city to a special administrative region of the People's Republic of China.[7]

Ho also identifies the subtle Lo Ting connections with the mermaids in Stephen Chow's film, and regards it as an attempt in allegorizing postcolonial Hong Kong's cultural identity.[8] In the film, while Shan may have the appearance of a beautiful mermaid in Hans Christian Andersen's fable, the mystical species are by no means just a Chinese replica of Western mermaids. In fact, they have more in common with the legendary ancestors of Hong Kong's boat people, the Lo Ting tribe, than with their western counterparts. When Liu Xuan first discovers Shan and her family's true identity, he runs to the police station and reports about being kidnapped by mermaids. When the police officers could not make out of what he describes as mermaids, they

Figure 11.1   Art exhibit "Lo Ting, Children of the Rebel", in *Hong Kong Reincarnated: New Lo Ting Archeological Find* curated by Oscar Ho. Hong Kong Arts Centre. 1997. *Source*: Oscar Ho.

draw a number of peculiar looking sketches of half-human half-fish creatures for him to identify, which invariably resemble the images of Lo Ting. Although the film has not made it explicit, Shan and her tribe's marginalized existence at the border of land and water, echoes how the Lo Ting species were depicted in the folklores. "By situating Lo Ting against the backdrop of Hong Kong's handover to China," these artistic representations of the legendary Lo Ting and their history, facilitate in a multispecies world-making activity that explores "the credibility of the official narrative of Hong Kong history and as to whether there is a unique Hong Kong identity to be cultivated and preserved." (Ho 1998) Such a discourse of endangerment, which reflects the filmmakers and art workers' imaginations of species-crossing based on references to traditional indigenous folklores, inform us to find better ways to develop toward what Heise terms "multispecies justice and multispecies cosmopolitanism." (Heise 2016: 6) Working on western indigenous literature like Linda Hogan's novel *People of the Whale* and Ihimaera's *The Whale Rider*, Adamson examines how stories about transformational animals, such as whales, bears, snakes, frogs, with their quasi-human qualities, "suggest new modes of research being referred to as 'multi-species ethnography' by scholars who study humans within the 'cosmos' of their entanglements with other kinds of living beings." (Adamson 29) She argues these indigenous works "illustrate that characters with transformational qualities are often employed to address social and environmental injustices by suggesting that 'boundaries are permeable'" and they "encourage readers to think about how they might alter 'the power relations at the root of social and ecological problems." (Adamson 34)

Similar issues have been raised in Chinese context in *The Mermaid*. As narrated by the female elder of the mermaid tribe, in ancient time, both humans and mermaids were evolved from apes. The difference in the mermaids comes from their ancestors' extended lives in the water resulting in the evolution of body and tail for underwater movement. Since then, the two species have been in opposition. While symbolically it reflects the political tension between Hong Kong and China, through Liu and Shan's romance, the movie also highlights the two species are sharing the same origin and root, thence to promote multispecies compassion and peaceful coexistence between humans and nonhumans.

SPECIES ENDANGERMENT: THE
CHINESE WHITE DOLPHINS

While the Lo Ting creatures are regarded as the prototype of Chow's mermaids, there is also another layer to *The Mermaid* that makes it a multispecies allegory of Hong Kong. The endangered mermaids is also a reflection of another nonhuman species in today's Hong Kong, the increasingly endangered pink dolphins, more commonly named Chinese white dolphins, which inhabit the Pearl River Delta and are drastically reducing in number.

Anthropologist Tim Choy discusses what he calls the "ethnography of endangerment" in Hong Kong in relation to the pink dolphins in Sha Lo Wan and the fishing village Tai O, in which he explores "the technical and affective production of endangerment [. . .] that enable certain forms of life to [become] objects of knowledge and love, as endangered things." (Choy 27) Chinese white dolphins have been living in the waters of Hong Kong for hundreds of years. They are the indigenous residents of Hong Kong. Through his studies of Chinese white dolphins (中華白海豚) since the 1990s, Choy reflects on Hong Kong people's concern for ecological issues and the change in social values since the handover in 1997. Chinese white dolphins have been seen in the waters of Tai-O, some off-shore islands, and the Pearl River Estuary. Attention by environmentalists was recorded in the 1980s. Dubbed the "pandas of the sea" due to their rarity and friendliness, in 1997, Chinese white dolphins were chosen as the official handover mascot, of the return of Hong Kong's sovereignty to People's Republic of China.

In recent years, however, land reclamations and sea-water pollution have resulted in an increasing number of stranded dolphins on shore. In the first six months of 2016 alone, fifteen dolphins were found stranded (Ng 2016). According to the Agriculture, Fisheries and Conservation Department, the number of dolphins has fallen by 80 percent over the past decade and a half, down from 188 in 2003, to fewer than 50 spotted between 2017 and 2018. In December 2017, the IUCN Red list of Threatened Species Status of the pink dolphins as updated from "Near Threatened" to "Vulnerable." (WWF-Hong Kong 2017) According to a more recent government estimate dated April 2020, only thirty two Chinese white dolphins are left in Hong Kong's waters. (Hutton 2020; Low 2020; Yip 2020) Threats to and damages on dolphins' habitat have come from marine infrastructural construction projects namely the Hong Kong-Zhuhai-Macau Bridge, the Third Runway for Hong Kong International Airport and reclamation works for a man-made island. These engineering projects have also brought about serious air pollution to human habitats in the north-west areas of Lantau Island (WWF-Hong Kong 2014; Ives 2017).

Fiona Law suggests that the name and terrain of Green Gulf in *The Mermaid* could most likely have used Sha Lo Wan, a similar-looking bay area in the northwest Lantau Island, as its blueprint. (Law 84–85) In *The Mermaid*, when the business tycoon Liu Xuan purchases the Green Gulf for reclamation, he comes up with a plan to using sonar transmitter to exterminate and cast the dolphins away from the region, which is also a subtle critique to "the

local controversies about whether the Chinese White dolphins would temporarily move from their habitat until the construction project of the third [Hong
Kong International Airport] runaway is completed." (Law 85)

In this regard, the story background of deterritorialization in *The Mermaid*
and the hidden local attachment sentiment have demonstrated, on the one
hand, as a commercial comedy, its aim is to convey a direct environmentalist message to promote compassion toward animals and to preserve marine
ecosystem and, on the other, to audiences knowledgeable of Hong Kong's
history and marine culture, the film can simultaneously be seen as an ecological metaphor against Hong Kong's reclamation works and for the protection
of Chinese white dolphins and other marine lives. Based on the above argument, it would appear that *The Mermaid* has skillfully solved the dilemma in
the overemphasis in local consciousness at the neglect of more universalized
global perspectives, or vice versa.

While the multiple layers of ecocritical messages appear to be *The
Mermaid*'s strength, the film falls short in other aspects. In the next section,
our discussion will focus on the depictions of eco-empathy and multispecies
justice in *The Mermaid*, to point out that despite the intention to promote
cross-species ethics of care and compassion, the film fails to break away
from the inherent doctrines of anthropocentrism, thus drastically weakens the
environmental message it tries to convey.

ECO-EMPATHY AND MULTISPECIES JUSTICE

In recent years, literary and film scholars have begun to explore and theorize
the variety of empathies, particularly in the study of animals and multispecies
contexts. Gruen calls it an entangled empathy, which describes "[a]n experiential process involving a blend of emotion and cognition in which we recognize we are in relationships with others and are called upon to be responsive
and responsible in these relationships by attending to another's needs, interests, desires, vulnerabilities, hopes, and sensitivities." (Gruen 3) Others also
explore more generally the extent in which literary and cinematic texts could
promote species-crossing empathies that facilitate us to think, feel and act
more like other nonhuman species, and be physically and emotionally more
in synch with them. (Monani 2014; Weik von Mossner 2017; St Ours 2020)
In *The Mermaid*, we could see how different types of eco-empathy are being
represented, through the character Liu Xuan and his gradual transformation.

The Mermaid opens with a montage sequence of environmental destructions that anticipates viewers in the environmental focus of the film consisting
of real-life documentary footage of air and river pollution caused by industrial wastes, deforestation, birds covered in petroleum, and coastal massacre

of dolphins taken from Louie Psihoyos' award winning documentary *The Cove* (2009). The latter image appears again in a later scene when Liu realizes his development project is causing massive disaster and destruction to the sea, as well as the lives of sea creatures and the mermaid community. He becomes obsessed with searching online for images and videos related to how human beings are causing destruction to the oceans, which take viewers back to the images we saw in the opening sequence. These online videos and images become the first trigger that causes Liu to change his attitude and perspective toward the sea, forming his empathy for the nonhuman species living in it. According to Gruen's categorization of empathies, here Liu develops a "primary/ personal empathy" which, as an empathetic individual, he "is able to connect [his] feelings to the reality of the individual being empathized with [the sea animals that are slaughtered]" (Gruen 47). As Gruen describes it, "personal empathy allows one to empathize with the actual situation of another but does not distinguish one's own perspective from the perspective of the other. This is the type of empathy wherein an individual really puts herself in the shoes of the other and loses sight of herself." (ibid.) The scene is building a tripartite relationship among the suffering animals, Liu Xuan, and us (viewers of the movie). By showing us the footages Liu sees, and letting us witness his gradual change in attitude and perceptual transformation, the movie invites us to identify with Liu and urges us to share Liu's growing empathy toward the animals and the mermaids too.

A second trigger Liu receives in the film comes when he volunteers himself to enter the sonar simulating room, to receive first-hand experience of the pain and harm invoked by the sonar transmission. This could be seen as another type of empathy which Gruen terms "cognitive empathy", which consists of the empathizer "not mirroring or projecting onto the emotions of the one being empathized with, but is engaging in a reflective act of imagination that puts her into the object's situation and/or frame of mind, and allows her to take the perspective of the other." (Gruen 48) As Law suggests, "[t]he experience of situated understanding and learning to feel (instead of seeing) the pain of others is the critical moment of change in the plot. This scene does not only convert [Liu] from the villain to an eco-hero but shows that it is possible for a greedy human to become a caring agent." (Law 90)

While Law sees it as the turning point of the story, showing us how a human character transforms himself into an empathetically caring agent, it remains questionable whether the film has conveyed its message of multispecies coexistence as effectively as we wish it to be. I'd instead argue the film is using cross-species ethics of care as a façade to disguise its deep ideological roots in anthropocentrism and consumer culture that operates in the human world. The film fails to foster eco-empathy more effectively, for two major reasons. First, *The Mermaid* has substituted eco-empathy with romance and comic reliefs. The reason for the film's inability to fully reflect this concept is its inappropriate handling of the relationship between Liu and the mermaid tribe. As a romantic comedy, *The Mermaid* places great emphasis on the flourishing of romance between Liu and Shan, resulting in the communication between Liu and the mermaids being limited to his affection for Shan, while remaining generally hostile or indifferent with the rest of the mermaid community. By focusing on romance, instead of inviting us to empathize and show compassion more broadly to a different species that could more realistically reflect human–nonhuman relationships in real life situations, the film fails to inform us of the actual modes of interactions and respect we should cultivate in a world of multispecies coexistence. Also, although there is no lack of brutal scenes of humans exerting violence on the mermaids, they have not been used to deepen ecocritical reflections. As a whole, the film does tell us that we humans should protect and care for other living beings. On the contrary, as the comedy, the film encourages viewers to laugh at sights of physical harms to mermaids, such as the character Octopus, while disguising himself as a human to undergo an assassination mission on Liu, his tentacles are mistaken as food by the restaurant chefs and are chopped off for teppanyaki grilling. In another scene, in order to make her way into the human world, Shan had her tail split into two halves for disguise as legs to walk on land. Scenes after scenes, we see attacks, mutilations, and suffering of the mermaids serving the purposes of comic reliefs in the romantic comedy. Toward the end, the film features a large-scale bloody massacre of the mermaids to reflect human greed and brutality. However, with most of the killed mermaids being unnamed, cardboard-like characters, there is little room for viewers' to identify and feel for them. That greatly reduced the impact of, hence our critical reflections upon, the images of violence on the nonhuman species.

Second, *The Mermaid* substitutes eco-empathy with humans' sympathy toward the nonhumans, from a privileged position. To explain this further, I'd like to first introduce social anthropologist Aihwa Ong's concept of "flexible citizenship" in relation to the political identity of Hong Kong people. For decades, Hong Kong has been comfortable with its ambivalent, mobile, and flexible identity positioning. There has always been this sense of pragmatism which Hong Kong people embrace. For decades they have been retaining "a flexible notion of citizenship and sovereignty as strategies to accumulate capital and power." (Ong 6) Ong coins the term "flexible citizenship" which she defines as "the cultural logics of capitalist accumulation, travel, and displacement that induce subjects to respond fluidly and opportunistically to changing political-economic conditions." (ibid.) Written in 1999, Ong sees Hong Kong's mixed Chinese and western culture as an exemplification of such spirit. Being a metaphor of Hong Kong people, I argue the legendary amphibian Lo Ting also exemplifies the notion of flexible citizenship, adding to it a multispecies dimension. In *The Mermaid*, with the mermaids possessing special abilities to walk, swim and live freely between land and water, they too exhibit an imaginary form of "multispecies" flexible citizenship. However, seeing it through the perspective of environmental and multispecies justice, have humans and mermaids attained equal status? Have mermaids given their rights to exercise their "flexible citizenship"? Has the film effectively reflected human's genuine attitude and true nature toward other species?

With the fairy-tale-like happy ending, Liu successfully rescued Shan and defeated the human villains who try to wipe out the mermaid clan entirely. The mermaid tribe, having avoided the slaughter, returns to the sea to search for a new residing place. In order to protect the mermaids from being discovered by the humans, Liu assists in keeping them safe in the underwater world, and tries to convince his own people on the land that mermaids do not exist. Shan hides her mermaid identity, and lives a simple married life with Liu in a huge deserted seaside mansion. On occasions, they would still return to the underwater world to enjoy sights of beautiful marine life, of dolphins, fishes, and corals and the likes. This echoes Ursula Heise's criticism toward contemporary people's commodification of the natural world, as she states,

> Some distinctly modern forms of intimate acquaintance with nature—highly specialized hobbies such as bird-watching or orchid-collecting—depend precisely on their being leisure activities rather than existential necessities; and they are often quite far removed from any genuine ecological understanding, focusing as they do on one particular aspect of ecology rather than its systemic functioning. A sense of place and the knowledge that comes with it, in other words, is something that most people quite rightly perceive as a kind of hobby. (Heise 2008: 55)

To Liu, the value of the mermaid world is no different. It may at best be used as a holiday resort with no value of its own. Will the underwater world's long-term wellbeing and regulation be of anyone's concern? Will the mermaid tribe be able to find a sustainable and permanent residing place underwater that can be safe from discovery and risk of invasion by humans? Will they still need to live in a threatened environment? How will the problem of pollution of the sea and ocean be resolved? Will there be a day when mermaids and humans finally accept each other and establish a peaceful, multispecies coexistence? With its avoidance to answer these crucial questions, the worlds of humans and mermaids continue their binary existence and we return to our old anthropocentric viewpoint, and the continuation of environmental and multispecies injustice that places humans on top of the hierarchy.

Despite the attempt to present ecological awareness via representations of multispecies justice and ethics of care, *The Mermaid* fails to depart from the logic of global capitalism, and thus returns to the inherent exploitative anthropocentrism and reveals humans' limited capacity to imagine true multispecies justice. In the end, the mermaid tribe, which represents more broadly the nonhuman animal world, remains deprived as a lower and exploited species. Its flexibility of identity as a "multispecies cosmopolitan citizen" (Heise 2008: 59; Heise 2016: 6) is overshadowed by its marginality. Behind the coexistence of humans and mermaids, there is no real subjectivity offered to the animal–human hybrid. Instead, it is actually the male protagonist Liu, who gets the real power, flexibility and subjectivity. The mermaids remain a marginalized, endangered species.

Distinguishing empathy from sympathy, Gruen argues that the latter "has the potential for being condescending. Because one attempts to keep one's own attitudes, beliefs, even prejudices distinct from the other, one can sympathize with another when sympathy isn't called for." (Gruen 45) In *The Mermaid*, with the reassertions of the anthropocentric values and the multibillionaire human protagonist "owning it all" at the end, it is apparent that sustaining eco-empathy and treating other species equally is not the ultimate point the movie can make. Instead, they remain an object for human sympathy.

Through the film, Chow attempts to encourage people to "take a look from the view of other living species at what humans have done,"[9] and from which to establish interspecies identification. The central message of the film from Stephen Chow has been delivered by the protagonists Shan and Liu, "If there's not a drop of clean water and a breath of clean air left in this world, making more money doesn't save you from your journey to death." And yet, the film's ending has ironically reasserted the logic of global capitalism, as reflected in the mermaids and the underwater world continuing to be commodified for human consumption. The worlds of humans and mermaids remain in binary opposition.

CONCLUSION: A TIME TO HEAL

Through its multispecies imagination and implicit place-based references to Hong Kong's past and present enviro-political situations, *The Mermaid* succeeds as a creative blockbuster comedy, and proves the commercial cinema's ability in conveying environmental messages that reconcile the local (place-specific) and the planetary (universal) concerns. However, it is also argued that the film fails to transcend beyond anthropocentric thinking, hence has not given multispecies justice more critical reflections. With the emerging discourses of endangerment, of both the environment and of various nonhuman species, there is an urgent need to practice planetary healing.

From Stephen Chow's *The Mermaid*, we could identify three major forms of healing from our discussion. First of all, by returning to the multispecies legends of the past, such as the folk tales of Lo Ting, we are brought to awareness that our cultural environment and worldviews have been shaped and affected by nonhuman species (fish, animals, plants) since ancient time, but our closer affinity with them were severed by modern civilization. Reconnecting with our natural, interspecies traditions and beliefs could be a first step in healing our increasingly damaged relationship with the more-than-human world. Second, to cultivate better understanding and practice of cross-species empathy, would also be crucial. Through our discussion of *The Mermaid*, it is shown that empathy can easily be confused with other emotional reactions and affections, such as sympathy and romance. Cultivating genuine eco-empathy, or multispecies empathy, could thus be a second step toward planetary healing.

Since early 2020, the coronavirus pandemic has posed serious threats to the world. However, with global consumptions and touristic activities being drastically reduced, wildlife is making a comeback. According to a recent report, there are signs of increase of Chinese white dolphins over the past few months.[10] This reminds us of *The Mermaid*'s happy ending scene in which our transspecies couple Liu and Shan dive under the beautiful sea with the colorful fish and corals. Behind the scene, however, lies the logic of global capitalism, where human activities and exploitation of the natural world and other nonhuman species stop planetary healing from happening. The last and perhaps most effective way of planetary healing could therefore be to resist such anthropocentric logic of global capitalism. In that way, our planet may have a chance to rejuvenate.

NOTES

1. "*The Mermaid*," (2016), Box Office Mojo https://www.boxofficemojo.com/movies/?id=mermaid2016.htm.

2. Hong Kong Film Critics Society, "Di Ershisan Jie Xianggang Dianying Pinglun Xuehui Dajiang dejiang liyou cuoyao," Hong Kong Film Critics Society. January 16, 2016. https://www.filmcritics.org.hk/%E5%AD%B8%E6%9C%83%E5%A4%A7%E7%8D%8E/%E5%BE%97%E7%8D%8E%E7%90%86%E7%94%B1/%E7%AC%AC%E4%BA%8C%E5%8D%81%E4%B8%89%E5%B1%86%E9%A6%99%E6%B8%AF%E9%9B%BB%E5%BD%B1%E8%A9%95%E8%AB%96%E5%AD%B8%E6%9C%83%E5%A4%A7%E7%8D%8E%E5%BE%97%E7%8D%8E%E7%90%86%E7%94%B1%E6%92%AE%E8%A6%81.

3. People's Daily, "Zhouxingchi pai meirenyu xiyin dafa Bazhuayu yijiao bei luozhixiang qiangzou," January 22, 2016. http://media.people.com.cn/n1/2016/0122/c14677-28075738.html.

4. Examples include the "Pig Sty Alley (zhulong chengzhai猪笼城寨)," in *Kung Fu Hustle* (with a play on its name of the demolished Kowloon Walled City), ancient settings in Journey to the West, and the fictitious Green Gulf and the mermaid world in *The Mermaid.*

5. These historical documentations, as Oscar Ho suggests, includes *New Sayings on Guangdong* (*Guangdong Xinyu* 廣東新語)*;* Fan Duangnang's *Guangdong Knowledge* (*Yue Zhong Jianwen* 粵中見聞)*;* John Richardson's *The Strange Tales of Guangdong* (*Guangdong Shuyi* 廣東述異)*;* and *Lingnan Narratives* (*Lingnan Congshu* 嶺南叢述).

6. The art exhibition "Hong Kong Reincarnated New Lo Ting Archeological Find" took place in Experimental Gallery, Hong Kong Arts Centre, Hong Kong between June 20 and July 24, 1998. https://aaa.org.hk/en/collection/event-database/hong-kong-reincarnated-new-lo-ting-archeological-find.

7. The cultural significance of Lo Ting to the re-imagination of Hong Kong identity is also reflected in the variety of creative texts ranging from literary writing to performance arts. In 2014, Hong Kong's Horizon Theatre produced a stage play Century's Dream of a Fishing Port using fishing port and Lo Ting as its theme. Inspired by the Lo Ting legends and influenced by a classic piece of work by the nineteenth-century British biologist Alfred R. Wallace's Hong Kong Mermaid , in which Wallace depicts his encounters with mermaids in Hong Kong, Japanese film director Iwai Shunji wrote his fantasy novel Wallace Mermaid (2013) which extended peoples' imagination of the legend of the mermaid.

8. "美人魚和盧亭：港人文化身份的隱喻," *Standnews.* June 22, 2016. https ://www.thestandnews.com/culture/%E7%BE%8E%E4%BA%BA%E9%AD%9A% E5%92%8C%E7%9B%A7%E4%BA%AD-%E6%B8%AF%E4%BA%BA%E6%9 6%87%E5%8C%96%E8%BA%AB%E4%BB%BD%E7%9A%84%E9%9A%B1% E5%96%BB/.

9. _____. "Zhouxingchi pai meirenyu xiyin dafa Bazhuayu yijiao bei luozhix-iang qiangzou," *China Daily Online.* January 22, 2016. http://media.people.com.cn/n1/2016/0122/c14677-28075738.html.

10. 勞敏儀"中華白海豚數量回升至52條　惟連續五年絕迹大嶼山東北水域," *HK01.* July 24, 2020. https://www.hk01.com/%E7%A4%BE%E6%9C%83%E6%96 %B0%E8%81%9E/502135/%E4%B8%AD%E8%8F%AF%E7%99%BD%E6%B5 %B7%E8%B1%9A%E6%95%B8%E9%87%8F%E5%9B%9E%E5%8D%87%E8% 87%B352%E6%A2%9D-%E6%83%9F%E9%80%A3%E7%BA%8C%E4%BA %94%E5%B9%B4%E7%B5%95%E8%BF%B9%E5%A4%A7%E5%B6%BC%E5 %B1%B1%E6%9D%B1%E5%8C%97%E6%B0%B4%E5%9F%9F.

WORKS CITED

Abbas, Ackbar. *Hong Kong: Culture and the Politics of Disappearance.* Hong Kong: Hong Kong University Press. 1997.

Adamson, Joni. "Whale as Cosmos: Multi-Species Ethnography and Contemporary Indigenous Cosmopolitics," *Revista Canaria de Estudios Ingleses*, 64, April 2012, pp. 29–45. ISSN: 0211-5913.

Adamson, Joni and Salma Monani. "Introduction: Cosmovisions, Ecocriticism, and Indigenous Studies," in *Ecocriticism and Indigenous Studies: Conversations from Earth to Cosmos*. Edited by Salma Monani and Joni Adamson. New York and London: Routledge. 2017, pp. 1–19.

Choy, Tim. *Ecologies of Comparison: An Ethnography of Endangerment in Hong Kong*. Durham and London: Duke University Press. 2011.

Gruen, Lori. *Entangled Empathy: An Alternative Ethics for Our Relationships with Animals*. New York: Lantern Books. 2015.

Heise, Ursula K. *Sense of Place and Sense of Planet: The Environmental Imagination of the Global*. Oxford and New York: Oxford University Press, 2008.

———. *Imagining Extinction: The Cultural Meanings of Endangered Species*. Chicago; London: University of Chicago Press. 2016.

Ho, Hing Kay Oscar. "Hong Kong Reincarnated New Lo Ting Archeological Find 香港三世書之再世書盧亭考古新發現," *Asia Art Archive Event Database*. 1998. https://aaa.org.hk/en/collection/search/library/hong-kong-reincarnated-new-lo-ting -archeological-find.

———. "The History of Lo Ting," in *Driving Lantau: Whisper of an Island*. Edited by Lo Yin Shan and Anthony McHugh. Hong Kong: MCCM Creations. 2012, pp. 167–177.

———. "Meirenyu he Luting: Gangren wenhua shenfen de yinyu (美人魚和盧亭：港人文化身份的隱喻)," *Stand News Hong Kong*. February 22, 2016. https://www.thestandnews.com/culture/%E7%BE%8E%E4%BA%BA%E9 %AD%9A%E5%92%8C%E7%9B%A7%E4%BA%AD-%E6%B8%AF%E4%BA %BA%E6%96%87%E5%8C%96%E8%BA%AB%E4%BB%BD%E7%9A%84 %E9%9A%B1%E5%96%BB/.

Hong Kong Film Critics Society, "Di Ershisan Jie Xianggang Dianying Pinglun Xuehui Dajiang dejiang liyou cuoyao," *Hong Kong Film Critics Society*. January 16, 2016. https://www.filmcritics.org.hk/%E5%AD%B8%E6%9C%83%E5%A4 %A7%E7%8D%8E/%E5%BE%97%E7%8D%8E%E7%90%86%E7%94%B1/ %E7%AC%AC%E4%BA%8C%E5%8D%81%E4%B8%89%E5%B1%86%E9 %A6%99%E6%B8%AF%E9%9B%BB%E5%BD%B1%E8%A9%95%E8%AB %96%E5%AD%B8%E6%9C%83%E5%A4%A7%E7%8D%8E%E5%BE%97% E7%8D%8E%E7%90%86%E7%94%B1%E6%92%AE%E8%A6%81.

Ives, Mike. "Hong Kong Wants a 3rd Runway. Will Its Dolphins Pay the Price?" *New York Times*. June 14, 2017. https://www.nytimes.com/2017/06/14/world/asia/hong -kong-white-dolphins-airport.html.

Kirksey, S. Eben. "The Emergence of Multispecies Ethnography," *Cultural Anthropology*, 25, no. 4, 2010, pp. 545–576.

Kirksey, Eben, Craig Schuetze, and Stefan Helmreich. "Introduction: Tactics of Multispecies Ethnography," in *The Multispecies Salon*. Edited by Eben Kirksey. Durham and London: Duke University Press. 2014, pp. 1–24.

Law, Fiona. "Fabulating Animals-Human Affinity: Towards an Ethics of Care in Monster Hunt and Mermaid," *Journal of Chinese Cinemas*, 11, no. 1, 2017, pp. 69–95.

Liu, Waitong (廖偉棠), "Meirenyu: Tongzhougongji de guaika zuqun wuweizachen de Xianggang Yinyu," (《美人魚》：同舟共濟的怪咖族群五味雜陳的香港隱喻), *Cinezen*. February 16, 2016. http://www.cinezen.hk/?p=5572.

Low, Zoe. "Hong Kong Environmental Group Urges Authorities Across Pearl River Delta to Expand Protected Marine Habitats to Save Chinese White Dolphins," *South China Morning Post*. June 2, 2020. https://www.scmp.com/news/hong-kong/health-environment/article/3087236/green-group-urges-authorities-across-pearl-river.

Monani, Salma. "Evoking Sympathy and Empathy: The Ecological Indians and Indigenous Eco-activism", in *Moving Environments: Affect, Emotion, Ecology, and Film*. Edited by Alexa Weik von Mossner. Ontario: Wilfrid Laurier University Press. 2014, pp. 225–247.

Ong, Aihwa. *Flexible Citizenship: The Cultural Logics of Transnationality*. Duke University Press. 1999.

St. Ours, Kathryn. "How Mushy Are Mushers? A Study in (Narrative) Empathy," *ISLE: Interdisciplinary Studies in Literature and Environment*, 27, no. 1, Winter 2020, pp. 27–45.

Weik von Mossner, Alexa. *Affective Ecologies: Empathy, Emotion, and Environmental Narratives*. Columbus: The Ohio State University Press. 2017.

WWF-Hong Kong. "Doubts about Mitigation Measures Prompt Green Groups to Recommend that ACE Members Reject the Third Runway EIA," *WWF-Hong Kong*, Press Release. September 12, 2014. https://www.wwf.org.hk/news/press_release/?12040/.

WWF-Hong Kong. "Protecting Chinese White Dolphins," *WWF-Hong Kong*. 2017. https://www.wwf.org.hk/en/whatwedo/oceans/advocating_for_more_marine_protected_areas/protecting_chinese_white_dolphins/.

Part III

MYRIAD THERAPEUTIC LANDS

Chapter 12

Displacement and Restoration
A Therapeutic Landscape in *311 Revival*
Kathryn Yalan Chang*

Ever since March 11, 2011, Fukushima, in the Tōhoku area of Japan, which had been hit first by a magnitude 9 earthquake, then by tsunami from Hokkaido to Tōhoku, and then to the Kanto region, and finally by nuclear radiation leak from the nuclear power plant accident, changed from "heaven" to "hell," just as the notable Chernobyl exclusion zone ringing the bell to people around the world. The region around the Plant was evacuated immediately.[1] As the documentary *311 Revival* (2017) comments, "March 11, 2011, the second time in human history of a Level-7 Severe Nuclear Accident. Since then, the world began to have a nameless fear towards Fukushima, and even towards its residents" (Tsoi and Mak 0.02.55).[2]

As the Japanese government has lifted the evacuation orders in certain areas of Fukushima gradually, this chapter approaches the documentary *311 Revival*, produced by Horatio Tsoi and Martina Mak (Tsoi and Mak 2017), by shifting the emphasis on the postapocalyptic landscapes in Fukushima to the possibility of return, revival, and recovery in the affected areas. This essay, with the main focus on evacuees and returnees in the Fukushima areas, argues *311 Revival*, as eco-cinema, not only reveals the dark side of environmental damages in Fukushima as a radiation-inflicted zone, but also advocates and promotes an activist spirit by revealing how a civic society or local citizens is able to devote themselves when the scale of the disaster challenges the governmental management and the trauma of the victims has been gradually forgotten as time goes by.

After the "the triple disaster" happened in Fukushima, the Japanese government, the social media, the individual, film directors, novelists, poets, artists, etc., compete to tell its stories. In order to show how "paradise arises

* This essay was supported by the Ministry of Science and Technology of the Republic of China (under the grant number MOST 109-2410-H-143 -004 -H).

201

in hell," *311 Revival* presents several stories and narratives in which environmental evacuees and voluntary migrants who concern about the affected area
went back to the original exclusion zone and devote themselves to the place
by action. Stories of life in postapocalyptic landscapes of Fukushima in *311
Revival* entail activism, spirit, action, and resilience especially when many
evacuees still have a second thought of returning back to their hometown
which belongs to disaster-prone terrains in Fukushima.

In the context of eco-cinema studies, the film *311 Revival* is an exemplar
of eco-cinema with regard to the environmental activist spirits it contains
and the eco-consciousness it conveys. Eco-cinema, according to Paula
Willoquet-Marcondi in *Framing the World: Explorations in Ecocriticism and
Film* (Willoquet-Marcondi 2010), suggests that "certain independent lyrical
and activist documentaries—not commercial (i.e., Hollywood) films—may
be thought of as eco-cinema because they are the most capable of inspiring
progressive eco-political discourse and action among viewers" (Rust and
Monani 2017, 3). Likewise, in the introduction to *Eco-cinema Theory and
Practice* (2012), Stephen Rust and Salma Monani also evoke the importance
of activism through the genre of eco-cinema. As they write, "Cinematic texts,
with their audiovisual presentations of individuals and their habitats, affect
our imaginations of the world around us, and thus, potentially, our actions
towards this world" (2). In other words, eco-cinema concerns not only about
if certain environmental consciousness has been conveyed but also about how
the film is produced in an eco-cinematic way. Cinema is "a form of negotiation," as Rust and Monani further explain, "a mediation that is itself ecologically placed as it consumes the entangled world around it, and in turn, is itself
consumed" (1). If eco-cinema studies means to "critically interrogate cinema's ecological dimensions and their implications for us and the more than
human world in which we live" (2), then Horatio Tsoi and Martina Mak's
film *311 Revival* demonstrate the possibility of re-inhabiting the homeland
in a critically interrogative way in terms of the message they convey and the
way they produce the film.

311 Revival is an independent production, collaborated by a production
team, including the presenter, Clarisse Yeung, a Wanchai district councilor
in Hong Kong, an Ex-HKTV producer Horatio Tsoi, the script writer Real
Cheung, and the production assistant Martina Mak. This is not Tsoi's first
Fukushima film. After the director Tsoi produced a TV program "Diaspora:
on Fukushima" with a focus on how social workers from Hong Kong help
residents of Fukushima on the aspect of mental health counseling after the
disaster, with Cheung's fluency in Japanese, Tsoi decided to shoot a documentary on Fukushima disaster "without earning for [themselves]" (Tsoi).[3]
With no support and resources of the TV station, the production and circulation of this documentary completely relies on collaboration and community-
based funding to cover travel expenses and others, including "further post

production work, audio mixing and dubbing, releasing the show at various community venues, [and] DVD distribution" (Tsoi). On the website of the introduction to their documentary, Tsoi expresses the independent production film is "Open to any type or form of collaboration with anyone who can help find a venue for community release" in order to reach more people and public. Regarding the circulation of the film, they also encourage "Fund raising through free donation along with any release [that] would be appreciated for further postproduction work on this documentary" (Tsoi).

The collaboration among directors Horatio Tsoi and Martina Mak, presenter Clarisse Yeung, local residents, NGO volunteer radiation measuring group, TEPCO official interviewed for the film (Kempton) illustrate how much efforts they should take to present what matters to displaced Fukushima people and strive for survival for tomorrow. Six years after the triple disaster, the team spent two weeks walking in January 2017 through the coastal areas of Fukushima, visiting six places in Fukushima—Namie, Iitate, Katsurao, Tomioka, Odaka, and Haramachi, and documenting different stories narrated by the evacuees and returnees through interviews. The film with its focuses on "revival" and "rebirth" of the environmental evacuees in Fukushima who save "their own communities by civil means" explores questions, to quote Tsoi's words, "how did they manage to rebuild and restore their mind and heart?; did civil self-help initiatives restore only what was lost?; or could it resurrect the place into a better community?" (Tsoi).

DISPLACEMENT AND NOSTALGIA

In the Anthropocene, places like Chernobyl and Fukushima end up becoming "a world without us" in Alan Weisman's sense.[4] The nuclear radiation accident in the Fukushima Daiichi Nuclear Power Plant on March 11, 2011, led to more than thousands of Fukushima "refugees" to spread from their hometown to other places in the form of eco-displacement. Like Chernobyl, evacuees or refugees of Fukushima saw their world and their lives as having two time frames "before 3/11" and "after 3/11".[5] The Fukushima Daiichi nuclear accident could be construed as standing for the discrepancy between memory's image (before 3/11) and current actuality (after 3/11). The "utopia" in the sense that "we are free to live and act another way" (Solnit 2010, 7) has become a "dystopia" after 3/11 that they'd never dare to return regarding their health and wealth. As the Japanese government has gradually lifted the evacuation orders in some areas in Fukushima, the Fukushima "refugees" are urged to accept the cancellation of the compensation money and return to the "hometown" (Kageyama 2016). As *The Japan Times* (March 11, 2016) shows, "When housing aid ends in April 2017, people in apartments under the government program will have to start paying rent or move out" (Kageyama). It is time again for them to decide if they should remain evacuated or restart life by returning home like years ago when they left their hometown (planned or not).

Hundreds of thousands of residents in Fukushima became evacuees or environmental refugees after the day March 11, 2011. Fukushima, famous for its agriculture, turns out to be "paradise on fire." Many people's hometowns have been deformed into a postapocalyptic landscape: deserted streets, abandoned houses, ghost towns, and tons of large black bags of radioactive soil and debris put on the temporary storage sites, a landscape that is a constant reminder of tragedies. Since the evacuation order was issued by the Japanese government after the triple disaster, environmentally displaced people of the Tōhoku area whose livelihood were rendered unsustainable by natural- and man-made disaster have undergone forced migration and voluntarily becoming migrants.[6] Till now, according to the Japanese government's Reconstruction Agency's statistics in January 2019, eight years after Tōhoku disaster in 2011, "approximately 54,000 evacuees remain in temporary lodging, including around 5,000 in temporary prefabricated structures" ("The State of Recovery").[7]

Although the word "evacuees" may suggest that "there is a place to go back" because "the situation is temporary" (Kageyama), evacuees of the Tōhoku areas are environmental migrants, as the definition of "environmental migration" in the International Organization for Migration (IOM) shows,

> Environmental migrants are persons or groups of persons who, predominantly for reasons of sudden or progressive change in the environment that adversely affects their lives or living conditions, are obliged to leave their habitual homes, or choose to do so, either temporarily or permanently, and who move either within their country or abroad. (*Glossary on Migration* 33)

Whatever residents whose lives have been influenced by the triple disaster in the affected zone should be called "refugees" or "migrants," mostly referring to permanent relocation, or "evacuees," granting the possibility to return, these people experience displacement and dispossession because "their lives and livelihood are at risk to an extreme environmental hazard" (Adger, de Campos and Moretreus 30), not to mention receiving the compensation which constantly reminds "their identity as disaster victims" (Tsoi and Mak 0.58.33).

Like victims of Chernobyl accident in 1886 who exemplify the notions of "bare life" in George Agamben's *Homo Sacer* comprehensively,[8] evacuees of Tōhoku area "feel like refugees [themselves] although they live in one of the world's richest and most peaceful nations" (Kageyama). Evacuees who are dislocated from their hometown and relocated in other areas through the arrangement of the government are excluded and, simultaneously, included by way of their own exclusion in George Agamben's "inclusive exclusion" (8). As geographer Doreen Massey argues that the way we characterize places is never simple but "fundamentally political" ("Double Articulation: A Place in the World" 114). Fukushima landscape, for example, represents the way in which Fukushima evacuees have been signified and labeled by non-Fukushima evacuees based on their relationship with "radioactive nature." All "made in Fukushima" produces, even the survivors of Fukushima, have been denigrated due to "potential connection between place and social identity" (Ranghieri and Ishiwatari 333). Evacuees in transition shelters, for example, have been harassed and excluded by "graffiti painted on vehicles or public buildings saying 'evacuee, go home'" (333). Although they are still included in the Japanese society, the feeling of "out of place" is enhanced when evacuees have been regarded as *kimin*, used to refer to Japanese emigrants sent to South America, meaning "discarded" by society and nation (Kageyama).

The displaced evacuees in the Tōhoku area, according to surveys conducted by Reiko Hasegawa (2012), demonstrates that they are uncertain about whether having the opportunity to go back home, about both the length of waiting for back to their houses and the "length of the decontamination process" because the zones "have to be decontaminated to reduce radiation levels," some of them were forced "to leave their houses, their lands, their animals" by "kill[ing] their cattle and destroy[ing] their harvests" (qtd. in Crimella and Dagnan 40). As Hasegawa reminds, "the suicide rate among evacuees, especially among those in temporary housing, is not negligible" (40). People who decided to leave voluntarily "as a strategy of precaution" (40) were blamed for their cowardice, betrayal of their hometown—"anti-Fukushima," and of their country—"anti-Japan" (Crimella and Dagnan 42). Tensions and conflicts "between those who left and those who stayed" in the zones increase, as the presenter of the film, Yang, comments, "Once you've been labeled by others as a victim or refugee, to go to live elsewhere, to lose your old, warm house, to lose your friendly neighbors, all these things, the psychological burden is tremendous" (Tsoi and Mak 0.32.40—0.33.00). Naturally, the burden of loss intrigues a sense of nostalgia, a feeling of missing the past and being homesick.

In studies of displacement, the concept of nostalgia, implying an "aching to return home," could be considered with respect to twofold nostalgia, defined by Liddell and Scott in *A Greek-English Lexicon* as νόστος meaning "return home" and as ἄλγος connoting "pain, grief, sufferings" (qtd. in Rohl 6). Following this definition, evacuees' and returnees' feelings toward their hometowns in *311 Revival* evoke Svetlana Boym's distinction of nostalgia, which could be split into "restorative nostalgia" and "reflective nostalgia." The former emphasizes "*nostos* (returning home) and proposes to rebuild the lost home and patch up the memory gaps" (Boym 41). On the other hand, reflective nostalgia, according to Svetlana Boym, "dwells in *algia* (aching), in longing and loss, the imperfect process of remembrance" (41). Such nostalgia's dual means have been mutually engaged in by the filmmakers of *311 Revival*. Evacuees who are incapable of going home due to employment, education, and other factors and who still wander outside the hometown experience reflective nostalgia in Boym's sense.

The past eight years for evacuees from Fukushima, to paraphrase Alexander Wilson's discussion of "nostalgic landscapes," have been a time of "dislocation, of ruptured communities, and irreversibly altered landscape" (204). The landscape of Fukushima after 3/11 changes in terms of its demography, alterations in agricultural production based on food safety checks, the political reorganization of peoples and lands, and the changing relations between individuals, groups, and soils. A strong sense of place toward their hometown from a nostalgic yearning for the past will be enhanced after evacuees have experienced losing it. Geographer Yi-fu Tuan verifies this strong attachment to a place by emphasizing the features of agricultural lifestyle, "Rootedness in the soil and the growth of pious feeling toward it seem natural to sedentary agricultural peoples" (Tuan 2001: 156). It is understandable that people who have the profound sentiment for land have strong reaction against the prolonged evacuation although this "profound attachment to the homeland," according to Tuan, "appears to be a worldwide phenomenon. It is not limited to any particular culture and economy" (2001: 154).

Unlike the situation in the colonial or political displacement in which going back home is never fully possible, the displaced people in Fukushima presented by *311 Revival* could be categorized as staying put in the affected zone, never going back to the hometown (planned or forced), and coming back after leaving the area. Among six places that the filming crew of *311 Revival* have visited, some of them are categorized as "difficult to return" zones, while others are undergoing the lift of evacuation orders.

Areas affected by nuclear radiation currently are divided into three categories. "Yellow and green areas can be regarded as habitable zones. But pink is classified 'difficult-to-return zones.'" (Tsoi and Mak 1.16.50).[9] Namie, the town of which 70% was classified as "difficult to return" (Tsoi and Mak 0.22.55),[10] is the place where people have difficulty deciding whether to be back due to the insufficient infrastructure and still-demolished zones. In remembrance of his hometown Namie, Masaya Yano, a singer whose family "don't plan to come back," dedicates his songs and music to his hometown. Looking at the deserted streets in his hometown, Yano, admitting that he never paid much attention to the nuclear plant next to them, wonders what makes it so hard to put into words (Tsoi and Mak 0.24.41). Every note and every lyrics of his music records his reflective nostalgic feelings for such changes (Tsoi and Mak 0.26.53).

With restorative nostalgic feelings for the hometown Namie, Kurosaka Chishio, a culinary chef and owner of Seafood Restaurant, who learned cooking in Tokyo after graduation from senior high in Namie, returns back to open a seafood restaurant after 3/11 in order to reconstruct the place. Nevertheless, acknowledging having tough times running a business and feeling uncertain about the future, Chishio still summons more people to join him: "It would be good if more young people would come back. Everyone feels terrible. I came back. I hope everyone will also come back" (Tsoi and Mak 0.29.59). A place like Namie, which used to be home for both Yano and Chishio, turns out to be a hell from heaven. As Chishio expresses, "Namie really is a good place. From days of old, it's kept its rustic feel. It has a sense of being loved and cherished." (Tsoi and Mak 0.28.58).

Like Namie, Iitate, thirty-five kilometers from the nuclear power plant, has become desolate after 3/11.[11] The land of Iitate piled with radiative wastes reminds the evacuees of their "homeless" identity. As Yoichi Tao, president of Resurrection of Fukushima Association, explains that Iitate that "received no compensation from TEPCO" was the place people took refuge at first. Residents of Iitate never thought they would become refugees until one month after the disaster when the "Japanese government classify Iitate as an evacuation zone" (Tsoi and Mak 1.11.06). Like Chishio, Muneu Kanno, a resident of Iilate, grows nostalgic for the past too, "I lived here with seven family members. I always thought life here was the best. It's truly a great place to settle down. With the four seasons, you could grow excellent crops year-round. But because of the nuclear disaster, all these beautiful things disappeared" (Tsoi and Mak 1.11.40–1.11.52).

Even though the sense of nostalgia implies the memorable past for the evacuees, "there is no one single 'past' to point to anyway," cautions Massey ("Double Articulation: A Place in the World" 113). Likewise, Arif Dirlik's "Place-Based Imagination: Globalism and the Politics of Place" also shows, "under the regime of modernity, the defense of place was [. . .] to a nostalgic yearning for a past lost irrevocably" (21). What is more, evacuees' hometown attachment or patriotic sentiments represented in the film *311 Revival* has been accused just as a kind of empty slogan of "I love my hometown" and devotes nothing to reduce uncertainty and the fear of the nuclear radiation in Yen-min Liao's opinions (Liao). Liao's skepticism that shows hometown nostalgia have little incentive to summon people back may overlook the fact that "place" in the Fukushima accident evokes Massey's statement that shows "change is acknowledged" ("Double Articulation: A Place in the World" 111). The "dynamic" changing relations, which are represented in the way of "major shifts and restructurings" ("Double Articulation: A Place in the World" 115), for example, are very often to be seen in the age of Anthropocene, where natural and man-made disasters hugely influence these relations.

Comprehending that their home places will never be the same after 3/11 and the difficult-to-return zones have "police patrolling nearby, as though awaiting the residents' return" (Tsoi and Mak 1.25.18), Yoko Endo admits in *311 Revival*, "everybody feel very conflicted in their hearts, [. . .] [but] no one has given up their hometown. If possible, everyone wants to come back. But now, we cannot. So everyone made different decisions" (Tsoi and Mak 1.25.20–1.25.33). The narrator of the film notices that although what contribution each person makes is small, it's still worth doing "something to unite people spiritually, something that everyone can emotionally attach to, and thus create a sense of identity. When everyone works toward a common goal, working together to write a song, there's feeling of mutual encouragement" (Tsoi and Mak 0.31.38). Such mutual encouragement and engagement is the basic spirit that consists of a therapeutic landscape.

A THERAPEUTIC LANDSCAPE: ACTIVE ENGAGEMENT AND COMMUNAL POWER

Landscape, as Denis E. Cosgrove indicates, is "an ideological concept" (15). What the film *311 Revival* exhibits is not the land of devastation but shows how Fukushima is gradually transforming into a therapeutic landscape with some stories in which local people, events and activities are embedded in various locations of the disaster zones. As the director of *311 Revival* explains, "Our crew went all over the coastal areas of Fukushima, recording the stories of the residents, each finding their own ways to help themselves" (Tsoi). Therapeutic landscape, a term referring to health geography, was developed by Wilbert M. Gesler in 1992 to "examine the healing dimensions of specific sites" (Winchester and McGrath ii). *311 Revival* first entails the questions if Fukushima refugees are struggling to return and then explores the possibilities of revival and resurrection in the affected areas by weaving different stories of losses and regains in Fukushima.

Despite the affections that the documentary *311 Revival* attempts to convey by demonstrating the efforts that people in the affected zones on the road of resilience make, such concerns are relentlessly criticized by Liao, who condemns the film for its demand for representing the resurrection of Fukushima and who argues such issue is nothing but should be left to the Japanese government or the decision made by individual Fukushima refugees; we outsiders should not intervene or butt in (Laio). The critique made by Liao risks getting caught in the insider—outsider dilemma. Indeed, "The issue of return is highly sensitive and politicized" (Crimella and Dagnan 44), but Liao's insistence on leaving the issue to the "insiders" themselves may end up putting them in an isolated condition and fails to recognize every local one is originally an outsider (Massey, "Double Articulation: A Place in the World" 115). In Massey's discussion of the local in the global, she asserts, "the 'other' is already within"; that is, the global is "everywhere and already, in one way or another, implicated in the local" ("Double Articulation: A Place in the World" 120).

311 Revival evokes, following Alexa Weik von Mossner's observation on the audiovisual texts or eco-cinema, environments in the affected zones "in the embodied minds of viewers" or "cinematic environments," in Mossner's words, will "come to life for viewers on the emotional level" (14). In other words, the film production of *311 Revival* not only helps more people outside the Fukushima understand the dilemma that displacements of Fukushima refugees suffer and confront but also re-imagine the landscape as a therapeutic one, in which the communication is carried out by way of the singing, the cooking, the festive display of locally distinct features of these places, to activate and persist in a reciprocal process of "transportation and immersion" (Mossner 14), which indicates that moving images invite viewers to be emotionally engaged and, literally, to borrow Erin James's interpretation of storyworlds, "come to know what it is like to experience a space and time different from that of their [. . .] environment" (xi).

"Gastronomies of place" (Megan Kimble) serves as a way to gather a crowd and also a symbol of flourish to activate the sense of place again, which brings a feeling of emotional attachment, developed by "individuals and communities in the course of living and growing within the setting of home" (Rose 97). Even a small food cart in the affected area can turn the spot as a place where people nearby came together. Hanaoka Takayuki, a civil servant in Tokyo, sells coffee and tea in food cart every weekend and "successfully attracting nearby residents to come to rest and chat" (Tsoi and Mak 0.34.49). On the other hand, two-thirds of the residents in Kawauchi, a village, twenty kilometer from Fukushima Nuclear Power Plant, had returned because it "had the earliest plan of returning home after March 11" (Tsoi and Mak 1.01.28). In order to boost the local tourism, Kevin Vongvanich, originally a factory owner in Kawauchi, "got a franchise contract and chose to open shop at Kawauchi," the first franchise coffee shop from Thailand at the end of 2016. Vongvanich's shop at Kawauchi challenges a traditional view of place which is "inevitable tied up with backward-looking nostalgia, with stasis, and with reaction" (Massey, "Double Articulation: A Place in the World" 114). Rather than giving places "single, fixed identities and to define them as bounded and enclosed, characterized by their own internal history, and through their differentiation from 'outside'" (Massey, "Double Articulation:

A Place in the World" 114), Vongvanich's shop still recognizes and appreciates the "specificity and uniqueness" of a place ("Double Articulation: A Place in the World" 117) by not only using "many of Fukushima's wood and timber" in the interior, but also "gradually tak[ing] locally grown vegetables to integrate them into the menu to create a style unique to Kawauchi" (Tsoi and Mak 1.00.35—1.00.40). As the narrator of the film summarizes, "To allow a non-Japanese coffee shop franchise to operate in Kawauchi, whose main industry was agriculture, forestry, and sea produce, can be understood to be entirely for the sake of revival" (Tsoi and Mak 1.00.35—1.00.45).

311 Revival brings in more stories of how "outsiders" help assist the affected places in terms of recovery. Never is governmental effort enough, as Ranghieri and Ishiwatari indicate that "universities and civil society organizations [which] have provided additional support for the affected population" (334). The focus on the communal power in *311 Revival* also brings attention to break the "easy assumptions" between locality and community (Massey, "Double Articulation: A Place in the World" 110). As Massey writes, "Very often, when we think of what we mean by a *place*, we picture a settled community, a locality with a distinct character—physical, economic and cultural" ("The conceptualization of place" 46). The communal power that reconstruction and revitalization in the affected areas possess enhances the notion Massey celebrates, a notion of community "that is international, and political affiliations [which] may likewise engender feelings of commonality, of shared heritage, visions, and commitments, which do not depend upon place" ("Double Articulation: A Place in the World" 111). *311 Revival* has presented organizations that stand for "outsider within" in the process of revival and help constitute the therapeutic landscape.

Safecast, an international community that provides real-time environmental information about the status of the radiation, air quality, and open data, promotes citizen engagement and empowerment after the 3/11 triple disaster happens. As a global organization that has branches internationally, Safecast, whose Tokyo office encourages the public to make "their own radiation meter," even went to Fukushima "to measure radiation levels. Now, even government officials reference the radiation maps they made" (Tsoi and Mak 0.05.45–0.05.59). What most characterizes Safecast, in *311 Revival*'s view, is sharing of knowledge—a characteristic that confirms the public's rights to know and a "co-participation model" based on citizen science, in which they insist that "all collected data is published online. There are absolutely no secrets." The D.I.Y. radiation meters lay emphasis on the sense that what they are doing is mostly about "education and empowerment, not just environment and radiation" (Tsoi and Mak 0.07.45-0.07.48).

If Safecast advocates "participation and empowerment" (Tsoi and Mak 0.08.00) by promoting citizen science, Shuichi Suzuki uses citizen science as a judgment to return to his hometown two days after 3/11. Suzuki's personal acquisition of radiation knowledge exemplifies Safecast's stress on citizen science, which is the foundation knowledge of therapeutic landscape in Fukushima because any therapeutic elements will be based on a safe and secure sense and understanding. Suzuki, an owner of Ramenya SUZU in Haramachi, insists on continuing his ramen shop after 3/11, which he has operated for fifteen years. Upon knowing that people in Haramachi are "out of food and gas" in the cold winter, Suzuki evaluated the level of radiation on his own and "came back from the opposite direction to support the search and rescue teams" (Tsoi and Mak 0.42.33). Suzuki's privately driving back in the night is based on his knowledge of the radiation level, which "was not really that dangerous" for him (Tsoi and Mak 0.42.11). Suzuki's rational understanding of nuclear power in *311 Revival*, simultaneously, defies Liao's criticism which blames the film for transmitting false irrational fear of radiation.

Undeniably, possessing sentimental feelings toward the nuclear power disaster is at stake in endeavoring to create a false alarm or fall into ignorance of rational understanding and knowledge about nuclear power. Cognitive understanding of nuclear power is necessary but insufficient in terms of community-based rehabilitation and revival. The critique of local residents' passion and affection toward their land and places represented in *311 Revival* fails to see what this film's contribution from the perspective of the effects of affect and emotion in the eco-cinema, which serves as to explore "affect and emotion and their reference to human experience of and attitudes toward non-human nature" (1), as Alexa Weik von Mossner argues in *Affective Ecologies*. *311 Revival*, in this regard, not only exhibits the evacuees' emotion in relation to the human and more-than-human world, but also exhibits evacuees' anger, fear, and affection toward their home places. As Margaret Winchester and Janet McGrath indicate that the therapeutic landscape has "both physical and symbolic dimensions" (viii). Decontamination job in Fukushima worries workers about their radiation exposure, an anxiety represented in *311 Revival* that causes Liao's suspicion, which again condemns the film's inadequate investigation of nuclear radiation exposure. The skepticism overlooks the distinction between "expert and lay knowledge about risks and risky behavior and the ways in which knowledge is constructed, and meanings negotiated, in the context of people's everyday lives" (Abbott, Wallace, and Beck 108). To explore the sense of uneasiness and fear that workers who take part in radioactive decontamination work in Fukushima prefecture possess, in other words, leads to the understanding that "Individuals rarely make decisions based solely on professional knowledge" (108).

Suzuki's rational understanding of nuclear power plant nearby urges him to publicly suggest that every citizen in Japan, the government included, should be completely able to "be equipped with relevant knowledge" about nuclear power and other alternative energy just as the certain instructions would be applied when "seeking refuge in cases of earthquakes and tsunamis" (Tsoi and Mak 0.44.00—0.44.05). Like Suzuki, who cares about the alternative energy and complains that there is not enough discussion in Japan as time goes by, Yoko Endo, a retired professor and an organizer of Tomioka Reconstruct Solar Energy Plan, together with her family and other residents decide to build a large-scale solar panel facility on an irreversible land overgrown with weeds in Tomioka, a place only ten kilometers away from the nuclear power plant, by "using crowd funding" for nine billion Yen (Tsoi and Mak 1.20.51). With an ambition to propose a Japan's largest civic crowd-sourced solar power plan and to change the desolate landscape which is full of weeds into a "community solar" (Adamson), Endo believes if they put solar panels, "at least every year people will come to cut the weeds. They can also regularly check whether radiation levels from the soil [have] reduced" (Tsoi and Mak 1.22.09).[12] With the belief that [her] home can generate electricity even without the nuclear power plant (1.21.52), Endo installs "5.5 kw solar panels" (Tsoi and Mak 1.21.48) on her own roof. Determining to change the desolate landscape into a "therapeutic" landscape covered with solar panels, Endo believes that "In reality, to put the nuclear plant into permanent disuse will take thirty to fifty years. Does the next generation not move on? So, we should first try solar power" (Tsoi and Mak 1.24.00–1.24.49). She, already seventy years old now, will be almost ninety if the panel operates for twenty years, for Endo believes "this is for the Earth, for the people who live here" (Tsoi and Mak 1.23.25). Endo's reflective nostalgia toward her hometown transforms Tomioka into a place which is beneficial to the future generations.

Indeed, therapeutic landscape as a notion, which analyzes how place and well-being work together" and which emphasizes "healing process works itself out in places (or situations, locales, settings, and milieus)" (Gesler 743) is not for one generation only. What comprising the concept of the therapeutic landscape in *311 Revival*, which could be classified as filmed figures who are embodied through their corporeal presence in the radioactive zones and who are embedded on the grounds of the disaster-stricken communities through events and activities is in consideration of the sustainable development and the alternative view of the landscape.

Masami Yoshizawa and his cows are embodied through their corporeal presence in the radioactive zones. In the aftermath of the disaster, radiation in the air had a negative effect on the fauna and flora of the radioactively contaminated areas. In order to decontaminate all radioactive particles, the government implemented a program to kill doomed livestock, including cows in Namie, a radiation hotspot. Yoshizawa, born and raised in Namie and the owner of the Ranch of Hope, refusing to obey the evacuation order including extermination of all his cows that he raised for years, saves his cows from becoming an obsolete object to be destroyed. In order to preserve his "radiative cows" who have become "moving rubber" that have no economic values, Yoshizawa practices "civil disobedience" by driving to "TEPCO's headquarters to protest" (Tsoi and Mak 0.16.41). Yoshizawa's resistance against the government evokes Julia Kristeva's notion of the abject which has been excluded but which still resists "its master" (2). Being an abject other who "embarked on the road of resistance" brings Yoshizawa into closer proximity with worlds of his cows, Yoshizawa follows and extends Agamben's notion of "inclusive exclusion" of bare life in the way it "serves to include what is excluded" (20). By including nonhuman entities into the realm of Agamben's "inclusive exclusion" and with the hope that lets "these 400 cows live to old age," Yoshizawa has maintained the ranch's operation "through donations, while giving speeches both in and out of country" (Tsoi and Mak 0.18.30) over the past few years till now.[13]

Refusing to be muted or excluded by the government, Yoshizawa explains the reason of staying rebellion,

> People here are too obedient. If you obey, then you'll be quickly compensated. People only discuss how much compensation they get. That's all we discuss. Of course, compensation is our right. But this place has been hit with a big disaster. With Namie suffering such a situation, in order to regain what rightfully belongs to us, in order to stop nuclear power, we victims should tell the whole country what has happened over these past six years" (Tsoi and Mak 0.19.44).

Even the place is deserted as Iitate or Namie, there is still hope if the power of self-reliance exists. Yoshizawa's active engagement in the process of resurrection gives cows "new meaning." Naming his ranch a Ranch of Hope, Yoshizawa believes, "Hope is not something prepared for you. It needs you to think about it. Only you can create hope" (Tsoi and Mak 0.19.05–0.20.52). Regarding his "radiative cattle" as companion and friends (Tsoi and Mak 0.21.03), Yoshizawa and his community (cattle and supporters) insist that "To have a life that is cherished is a kind of happiness whether that life is of a cat, a dog, a pig, or a chicken" (Tsoi and Mak 0.21.26). Yoshizawa's insistence on citizen engagement about the nuclear accident includes the lives of animals.

The festive display of locally distinct features of the affected places makes the beautiful picture of the therapeutic landscape in Fukushima. Citizen engagement, shown in Yoshizawa and his community, has also achieved its effect on preserving horses and facilitating communal motives of affiliation through local customs and rituals in Haramachi of Minamisoma. A famous traditional activity "that draws countless foreign tourists every summer to observe" (0.12.35) is held every year in July at the Shatin Horseracing Tracks in Haramachi. People will dress up in traditional outfits with their horses in the ceremonies. As *311 Revival* reveals,

> After the March 11 earthquake, the government ordered the slaughter of all animals in the restricted zones. Basically none were spared. People from Minamisoma requested the government to spare the horses here because the horses' breed is very well known. After the horses escaped their death, very quickly this ceremony was held again, two years after the disaster. You see a cultural festival uniting the whole community (Tsoi and Mak 0.13.50).

As the film predicts that cultural festival is a force to unite the community and preserve its animals. The grand traditional Soma-Nomaoi festival that was held in Minamisoma in 2018 and that attends to samurai culture successfully engages local families to celebrate the history of their hometowns. Some of the places that *311 Revival* had visited belong to the areas that are in the process of lifting the ban of the evacuation zone. Art and festivals in Minamisoma that foreground spiritual dimensions and communal power prove that "Revival is not just about aspects of daily life. Spiritually, there needs to be some belief that unites everyone. This isn't something that some economic value, or the pursuit of economic value, can achieve" (Tsoi and Mak 0.13.40–0.13.52).

NGOs, which stand for another "outsider within" group, actively participate in the process of revival in Fukushima (Tsoi and Mak 1.03.28). As *311 Revival* points out, after 3/11 "non-governmental organizations began to appear" and they have helped "the ranchers here, and the residents, to rebuild their lives" (Tsoi and Mak 1.06.15–1.06.19). As many people used to live on agriculture in Fukushima, how to resume their formal farming life constitutes the priority of revival. Due to his confidence in the wasabi plant, Muneu Kanno, a sixty-six years old farmer from Iitate, goes back and grows crops again with NGO's assistance—the Resurrection of Fukushima Association. Kanno, under the guidance of NGO members, learns how to "use science to re-engage in crop growing," and it "becomes a testing point for Iitate's post-disaster redevelopment" (Tsoi and Mak 1.03.24).

Like Safecast, which advocates for citizen science, instead of relying completely on the government, the Resurrection of Fukushima Association as a NGO, founded by Yoichi Tao and established three months after the disaster, not only volunteers to "monitor radiation levels for the whole village," but also "take[s] advantage of their network to contact various universities, research centers, and business organizations" (Tsoi and Mak 1.07.50–1.08.10). With the data "published to the villagers and online" (Tsoi and Mak 1.08.15), the Association promotes citizen engagement and participation because "if you only ask the village government to be responsible for everything," continues Tao, the president of the Association, "true self-reliance and regeneration will never realize" (Tsoi and Mak 1.09.20).

The Association as an NGO also plays a pivotal role in recomposing the local residents' sense of place and "ecological identity" (Thomashow). As food safety is always the top concern within or outside Fukushima, the Association based on its recognition of the major industry in Iitate—agriculture, forestry, and farming, help conduct "all kinds of tests and experiments," finally successfully grow crops and rice in the fifth time, and make sure of "the radiation levels and evidence of its safe, gradual decline" (Tsoi and Mak 1.06.50–1.07.45). In other words, the Association resumes a "continuing presence of the sustaining group" (Sopher 136) because the sense of being "at-homeness," including the characteristics of "rootedness, appropriation, regeneration, at-easeness and warmth" (Seamon 78), and their previous "relations with their "nurturing and sheltering group[s] as they are associated with the landscape that gives it meaning as the landscape of home" (Sopher 136) have all lost after 3/11 disaster. Though still having no chance to sell any produces, the Association brings to light what Mitchell Thomashow calls "ecological identity" in which "ecosystem health, community well-being, and personal happiness" (143) are getting to restore. Bestowing a new meaning upon their "wounded ecosystem" encourages a new sense of place for residents, which comprises a therapeutic landscape that Winchester and McGrath propose: "When considering health, space is important not just in terms of exposure to risk of disease but also in terms of ability to achieve a therapeutic outcome" (viii). Restarting farming reconfigures the life in the affected zone and has "made the psychological burden lighter" (Tsoi and Mak 1.10.28)

Insisting on working "in a symbiotic mode" (Tsoi and Mak 1.09.24), the goal of NGO is to "achieve self-reliance and rebirth" (Tsoi and Mak 1.12.30). Emphasizing citizen engagement as a way to revival, Tao states,

> Rather than a volunteer group, you may call them partners on Iitate's path to revival. Farmers, such as Mr. Muneu Kanno, have also been directors at the Resurrection of Fukushima Association. Many residents, university professors, experts have become one of us. There are also others with jobs who come to help after work. We've gathered all kinds of people. (Tsoi and Mak 1.13.15)

As Yeung, the spokesperson of *311 Revival*, also summarizes, NGOs "are the treasure of each place" (Tsoi and Mak 1.14.35).

Nevertheless, those who have a strong feeling of restorative nostalgia return or remain in the affected zones with different views of re-building the communities. How to reach consensus among communities on recovery and rehabilitation plans is a process of negotiation. Imagining there is a community with no conflicts is a fantasy. As Chantal Mouffe indicates,

> Many communitarians seem to believe that we belong to only one community, defined empirically and even geographically, and that this community could be unified by a single idea of the common good. But we are in fact always multiple and contradictory subjects, inhabitants of a diversity of communities (as many, really, as the social relations in which we participate and the subject-position they define), constructed by a variety of discourses and precariously and temporarily sutured at the intersection of those positions. (44)

Although every group differs in terms of its views, members' backgrounds, and perspectives, as Massey suggests, "all of this is always in process, as new layers of meetings and understandings" (1994: 117). Every affected area in Fukushima is moving toward this new understanding with the spirits of citizen engagement and communal power to comprise the possible therapeutic landscape in Fukushima.

In the Anthropocene, (environmental) refugees are formed involuntarily. "Rootedness in a locality" or "an emotional commitment to it," as Tuan (1975: 152) indicates, are "increasingly rare," even though they desire to stay as the case in the Fukushima accident. Environmental refugees or evacuees in Tōhoku area are still suffering from the bitterness of the accident till now as *311 Revival* reveals although as time goes by people can easily "tune out and ignore or avoid the negative images, rationalizing their inaction, practicing denial and apathy" (Mitchell 143). An eco-cinema, like *311 Revival*, is able to provide a chance for people outside the affected area to have a cognitive and affective understanding of the area and its people. *311 Revival* offers an alternative vision of examining how Fukushima could be transformed from the land of devastation to the therapeutic landscape, which is intertwined with stories demonstrating examples of personal and community healing. Till now, Fukushima still needs not only the responsibility of the authorities but also from the citizens and the groups which are able to enact the power of transformation on the road of resilience.

NOTES

1. However, according to the survey in *The State of Environmental Migration 2011*, without expecting the second wave to hit, "the survivors have complained that the evacuation orders were not transmitted appropriately for the gravity of the

situation" and "the Japanese authorities admitted that the nuclear disaster found them unprepared" (Crimella and Dagnan 37).

2. As a matter of fact, Fukushima is one of the three most affected prefectures. The other two are Iwate and Miyagi. (Crimella and Dagnan 36).

3. Without steady income and with self-fund the shooting, Tsoi and Cheung finally have the assistance from an ex-colleague Mak, who took leave from her job and joined them, and Yeung, who agreed to help "pro bono" (Tsoi) and "her view became the subjective voice in presenting the people, circumstances and thoughts in the Fukushima disaster area as she saw it" (Tsoi).

4. Alan Weisman published a book, *A World without Us* (2007), in which he explores the impact that human activities have on the Earth.

5. This idea is inspired by Abbott, Wallace, and Beck's referring to "before Chernobyl" and "after Chernobyl" when "Chernobyl marked a radical break in their [victims'] lives and, in that sense, represented a genuine biographical disruption" (111).

6. Clara Crimella and Claire-Sophie Dagnan in *The State of Environmental Migration 2011* have shown that the statistics of "the displaced populations" intrigued by "the triple disaster" two days after the disasters caused "more than 450,000 evacuees from the tsunami and 170,000 related to the nuclear accident" (38).

7. *311 Revival* has the population statistics of the evacuees and returnees in the places that they have visited in 2016. In Haramachi, Fukushima, Population before 311: 47,116; returnees: 43,503. In Katsurao, Fukushima, population before 311: 1,567; evacuees now: 1,389. In Tomioka, Fukushima, population before 311: 15,960; evacuees now: 14,999.

8. See Davies, Thom. "A visual geography of Chernobyl: double exposure." International Labor and Working-class History, vol. 84, 2013. pp. 116–139. Thom Davies a, and Abel Polese. "Informality and Survival in Ukraine's Nuclear Landscape: Living with the Risks of Chernobyl." *Journal of Eurasian Studies* 6 (2015): 34–45.

9. According to the film's statistics, Namie's population before 3/11 was 21,434. At the time (January 2017) when the crew filmed in the area, it still has 20,833 evacuees outside of Namie.

10. Namie and Tomioka are municipalities that people cannot return among the four. Together with the other two municipalities—Okuma and Futaba, these four municipalities formulated the recovery plans from September 2012 to March 2013 although the plans "cover relocation to other municipalities, but they do not include detailed rehabilitation of the original communities" (Ranghieri and Ishiwatari 335). The basic concept that recovery plans include is as follows. "(1) build a safe, secure, and sustainable society free of nuclear power; (2) revitalize Fukushima by bringing together everyone who loves and cares about it; and (3) rehabilitate towns so they can be a source of pride again" (Ranghieri and Ishiwatari 334-335).

11. As *311 Revival* indicates, "Iitate is nearly 1/4 as big as Hong Kong. There originally lived 6,000 people here. Now, almost half have moved and fled to Fukushima City. About 700 still live in temporary housing areas. If you look farther ahead, that originally should be farmland and pastures. Now, they're piled with radiative soil from residential areas and publish facilities. Added up, they account for about

twenty-two million cubic meters, equivalent to 11,600 Olympic-sized swimming pools." (Tsoi and Mak 1.04.38).

12. "Even so, Mrs. Endo, who has political experience, still believes that this is the best way out for Tomioka. She and her family actively reach out to people from Tomioka scattered east and west, hoping that this huge plan can start as soon as possible. She expects this plan to begin in April 2019, and operate for twenty years. The target is to generate thirty-three megawatts of power, equivalent to 10,000 units of ordinary residential electricity consumption" (Tsoi and Mak 1.24.04). In 2039, Mrs. Endo will be eighty-nine years old. Who knows what will Tomioka be like then?" (Tsoi and Mak 1.24.13).

13. Yoshizawa ran for the election of town leader in Namie in 2018. Although having presumably lost the election, he never stops fighting against the government. https://www.kahoku.co.jp/tohokunews/201808/20180806_61026.html.

WORKS CITED

Abbott, Pamela, Claire Wallace, and Matthias Beck. "Chernobyl: Living with Risk and Uncertainty." *Health, Risk and Society* 8.2 (June 2006): 105–121. Print.

Adger, W. Neil, Ricardo Safra de Campos and Colette Moretreus. "Mobility, Displacement and Migration, and Their Interactions with Vulnerability and Adaptation to Environmental Risks." *Routledge Handbook of Environmental Displacement and Migration*. Eds. Robert McLeman and François Gemenne. New York: Routledge, 2018. 29–41. Print.

Agamben, George. *Home Sacer: Sovereign Power and Bare Life*. Trans. Daniel Heller-Roazen. Standford: Standford University Press, 1998. Print.

Boym, Svetlana. *The Future of Nostalgia*. New York: Basic Books, 2001. Print.

Cosgrove, Denis E. *Social Formation and Symbolic Landscape*. Madison, WI: The Uninversity of Wisconsin Press, 1988. Print.

Crimella, Clara and Claire-Sophie Dagnan. "The 11 March Triple Disaster in Japan." *The State of Environmental Migration 2011*. Ed. Francois Gemenne, Pauline Brucker, and Dina Ionesco. International Organization for Migration (IOM), 2012. 35–46. Print.

Dirlik, Arif. "Place-Based Imagination: Globalism and the Politics of Place." *Places and Politics in an Age of Globalization*. Ed. Roxann Prazniak and Arif Dirlik. New York: Rowman and Littlefield, 2001. 15–52. Print.

Gesler, Wilbert M. "Therapeutic Landscapes: Medical Issues in Light of the New Cultural Geography." *Social Science & Medicine* 34.7 (1992): 735–746. Print.

Glossary on Migration, 2nd Edition. International Migration Law No. 25, IOM, Geneva. 2011. n.d. Web. December 23, 2018. http://publications.iom.int/bookstore/index.php?main_page=product_info&cPath=56&products_id=1380.

James, Erin. *The Storyworld Accord: Econarratology and Postcolonial Narratives*. Lincoln: University of Nebraska Press, 2015. Print.

Kageyama, Yuri. "Nuclear Refugees Tell of Distrust, Pressure to Return to Fukushima." *The Japan Times*. March 11, 2016. Web. December 12, 2018.

Liao, Yen-min. "Interpreting *311 Revival*: True or False." March 20, 2017. Web. October 8, 2018. https://thestandnews.com/cosmos/%E8%A7%A3%E8%AE%80 -311%E5%BE%A9%E8%88%88%E8%88%87%E5%86%8D%E7%94%9F-%E7 %9A%84%E7%9C%9F%E7%9B%B8%E8%88%87%E7%9B%B2%E9%BB %9E/.

Liddell, H. G. and Scott, R. *A Greek-English Lexicon*. Oxford: Clarendon Press, 1940. Print.

Massey, Doreen. "The Conceptualization of Place." *A Place in the World*. Ed. Massey, D. and Jess, P. Oxford: Oxford University Press, 1995. 45–86. Print.

———. "Double Articulation: A Place in the World." *Displacement: Cultural Identities in Question*. Ed. Angelik Bammer. Bloomington: Indiana University Press, 1994. 110–121. Print.

Mouffe, Chantal. "Radical Democracy: Modern or Postmodern?" *Universal Abandon?: The Politics of Postmodernism*. Ed. Andrew Ross. Minneapolis: University of Minnesota Press, 1988. 31–45. Print.

Ranghieri, Federica and Mikio Ishiwatari, eds. *Learning from Magadisasters: Lessons from the Great East Japan Earthquake*. Washington, DC: The World Bank, 2014. Print.

Rohl, Darrell J. "Place Theory, Genealogy, and the Cultural Biography of Roman Monuments." TRAC 2014: Proceedings of the Twenty-Fourth Theoretical Roman Archaeology Conference, Reading, 2014. Oxford: Oxbow. Brindle, T., Allen, M., Durham, E., and Smith, A. Ed. 2015. 1–16. Print.

Rose, G. "Place and Identity: A Sense of Place." *A Place in the World*. Ed. Massey, D. and Jess, P. Oxford: Oxford University Press, 1995. 87–174. Print.

Rust, Stephen and Salma Monani. "Introduction: Cuts to Dissolves—Defining and Situating Ecocinema Studies." *Ecocinema Theory and Practice*. Ed. Stephen Rust, Salma Monani, and Sean Cubitt. New York: Routledge, 2013. 1–14. Print.

Safecast. Shuttleworth Foundation. Web. November 10, 2018. https://blog.safecast.org/.

Seamon, David. *A Geography of the Lifeworld: Movement, Rest and Encounter*. London: Croom Helm, 1979. Print.

Solnit, Rebecca. *A Paradise Built in Hell: The Extraordinary Communities that Arise in Disaster*. New York: Penguin Publishing Group, 2010. Print.

Sopher, David E. "The Landscape of Home: Myth, Experience, Social Meaning." *The Interpretation of Ordinary Landscapes*. Ed. Meinig, D. W. New York: Oxford University Press, 1976. 129–149. Print.

"The State of Recovery in Tōhoku Eight Years after 3/11." *Nippon.com*. n.d. March 7, 2019. Web. May 15, 2019.

Thomashow, Mitchell. *Ecological Identity: Becoming a Reflective Environmentalist*. Cambridge, MA: MIT Press, 2002. (1996). Print.

Tsoi, Horatio. *311 Revival* (311 復興與再生). COLLACTION. 2017. Web. September 15, 2018. https://www.collaction.hk/s/311revival?lang=en.

Tsoi, Horatio and Martina Mak. Dir. *311 Revival*. (311 復興與再生). 2017. DVD.

Tuan, Yi-fu. "Place: An Experiential Perspective." *Geographical Review* 65 (1975): 151–165. Print.

———. *Space and Place: The Perspective of Experience*. Minneapolis: University of Minnesota Press, 2001. Print.

Weik von Mossner, Alexa. *Affective Ecologies: Empathy, Emotion, and Environmental Narrative*. Cognitive Approaches to Culture. Columbus, OH: Ohio State University Press, 2017. Print.

Wilson, Alexander. *The Culture of Nature: North American Landscape from Disney to the Exxon Valdez*. Cambridge, MA: Blackwell, 1992. Print.

Winchester, Margaret and Janet McGrath. "Therapeutic Landscapes: Anthropological Perspectives on Health and Place." *MAT* (*Medicine Anthropology Theory*) 4.1 (2017): i–x. Print.

Chapter 13

Nuclear Power Plants, East Asia, and Planetary Healing

Philip F. Williams

INTRODUCTION

Carbon-free renewables such as solar energy panels and wind turbines are on a trajectory to contribute a steadily increasing proportion of overall electric generation in East Asia and elsewhere in the world. Any government subsidies still being directed to carbon-emitting fossil fuels should be severely reduced if not eliminated and redirected to such relatively carbon-free vehicles of electricity generation, preferably with the aid of a carbon tax (Ellis et al.). However, what can be relied upon to generate a constant base load of electricity all twenty-four hours of each day, including those extensive periods when the sun is not shining or the wind is not blowing? The default solution for most countries other than France and Lithuania, which for decades have derived at least three-quarters of their electricity through nuclear power plants with an excellent safety record, has been mostly greenhouse gas-emitting coal-fired, oil-fired, or natural gas-fired electric generating facilities.

Yet this default solution of relying upon fossil fuels for baseload electricity generation is increasingly reckless in the context of accelerating greenhouse gas emissions and climate change, especially for developing countries that have an underdeveloped nuclear power sector. As the Pulitzer Prize-winning journalist Richard Rhodes has pointed out, "Without nuclear power, the doubling of demand [for electricity] projected for the developing world in the next thirty years will be met mainly through coal—or at best natural gas, which produces fully half as much carbon dioxide as coal when it burns" (Rhodes 2019: 314).

Veteran US ecologist and pragmatic futurist Stewart Brand has relied on evidence-based scientific research into affordable, sustainable, and technologically grounded reductions of greenhouse gas emissions in writings such as *Whole Earth Discipline: Why Dense Cities, Nuclear Power, Transgenic Crops, Restored Wildlands, and Geoengineering Are Necessary* (2009). This chapter focuses on eco-pragmatist views on issues such as relying upon increasingly safe and passively cooled nuclear energy plants as a necessary base-load provider to complement relatively intermittent electricity suppliers from renewables like solar, wind, geothermal, and hydropower (Oberhaus 2020). This combination of energy sources could gradually help wean East Asia and the rest of the world from fossil fuels, especially the dirtiest and most polluting types like coal-fired power plants, and their climatically dangerous greenhouse gas emissions.

Headline-grabbing nuclear power plant accidents at Chernobyl in 1986 and Fukushima 福島 in 2011 have tended to obscure the fact that newer third-generation and fourth-generation nuclear plant technology prevents meltdowns and other accidents with a passive cooling design that requires no cooling-water pumps—and thus is effective even during electric power outages (Oberhaus 2020). Even with the first couple of generations of nuclear plants, no deaths or even life-threatening radiation exposures to humans have been confirmed from radiation leaks caused by accidents at Fukushima and Three Mile Island. For example, the worst accident at Chernobyl led to no more than a few dozen short-term radiation-related fatalities among fire fighters and other first responders and an estimated 3 percent rise in thyroid cancers, totaling an estimated 4,000 cancer fatalities out of a nearby population of 600,000 who were significantly exposed to radiation from this accident (Brand 92; Rhodes 2018). However regrettable any deaths from this worst of all nuclear accidents are, the number of fatalities from the Chernobyl accident pales in contrast with the estimated 171,000 Chinese persons who died in August 1975 after the Banqiao 板橋 and Shimantan 石漫灘 hydroelectric dams in Henan province broke apart during heavy rains from Typhoon Nina (Yi 25-38)—or the hundreds of thousands of people worldwide who die each year from respiratory and cardiac illnesses triggered by air pollutants emitted from fossil fuels burned in power plants, stoves, and internal combustion engines (Rhodes 2019).

Indeed, the safety record for even older nuclear plant designs has been very good overall, especially if compared with the dangerous effects of natural gas explosions, mercury-laden coal ash heaps, and deadly particulates and carbon dioxide emissions from fossil fuel combustion. The Japanese government has thus initiated plans to gradually restart most of its temporarily shut down nuclear power plants in an effort to eventually achieve roughly 20 percent of the country's electricity generation through its nuclear plants—down from over 30 percent prior to the Fukushima accident. This one-fifth component of nuclear energy is similar to the long-standing level of contribution of nuclear power plants to the United States' and Russia's total amount of electric generation, and meets the initial minimal target of 18 percent for the share of the nuclear power sector in a given country that was set by world-renowned scientists on the Intergovernmental Panel on Climate Change (IPCC) (Conca 2019).

The PRC government began to embark upon nuclear-powered electricity generation much later than such nuclear power stalwarts as France, the United States, the U.K., Russia, Lithuania, Finland, Japan, and Taiwan. As late as the mid-1980s, the PRC stood alone as the world's only nuclear-armed nation that entirely lacked nuclear power plants (Xu 16). Even South Korea brought its first nuclear power plant on line in 1978—roughly a decade and a half sooner than the PRC's first nuclear power plants came online in Daya Bay 大亞灣, Guangdong and Qinshan 秦山, Zhejiang in the early-to-mid 1990s. When the PRC began to plan to finance and build its first two nuclear power plants early in the post-Mao Reform Era during the early-to-mid 1980s, Taiwan already had two such nuclear power plants up and running, Jinshan 金山 and Kuosheng 國聖 [Guosheng]. Nonetheless, during the post-Mao era the PRC strove to catch up with its more technologically advanced neighbors, embarking upon an ambitious nationwide program of building up its nuclear power sector with a combination of imported and indigenous nuclear technology in order to reduce its lopsided dependence upon heavily polluting coal-fired facilities. This continuing overreliance upon coal for both heating and electric generation earned the PRC the dubious distinction of having twenty out of the world's thirty worst-polluted cities by the year 2000 (Ebrey 2010: 349).

Battery storage technology and more advanced efficiencies in solar photovoltaics and wind turbines might improve in the distant future to the point where renewables like wind and solar alone could supply electricity twenty-four hours per day and seven days per week to a wired-in and high-tech world that eschews most fossil fuels. However, as the IPCC has soberly pointed out, during the intervening decades between now and then, an eco-pragmatic approach that includes nuclear power plants of increasingly safe and efficient design producing roughly 18 percent of the planet's total electricity is arguably necessary for humankind to make sorely needed drastic cuts in fossil fuel combustion and greenhouse-gas emissions throughout East Asia and elsewhere in the world (Brand 2009: 88).

ISSUES OF SAFETY AND NONPROLIFERATION IN NUCLEAR POWER PLANTS

No form of electricity generation is risk-free for the health of humans and ecosystems. According to a scientific study of the health effects of five major forms of European electricity generation in *Lancet*, coal-fired plants have had the worst record in deaths from both accidents and air pollution health, while nuclear plants have had the best record by both measures (Markandya and Wilkinson 2007: 979–990; see the chart entitled "Nuclear is safest"). As illustrated by the eco-pragmatist documentary film *Planet of the Humans*, the industrial processes required to manufacture solar panels and wind turbines are themselves quite carbon-intensive and include many toxic components—while the shelf life of these devices is invariably limited and thus entails issues related to the disposal of worn-out equipment and its replacement with newly manufactured equipment (Gibbs 2020) (Figure 13.1).

Where nuclear energy stands out in the area of popular perception is the degree to which its risks have been grossly exaggerated by many environmentalists and other lay activists who know little or nothing about nuclear physics, nuclear engineering, and even cost–benefit analysis. As Brand has noted about widespread yet scientifically baseless phobias about transgenic crops or "GMOs," the more evidence-based knowledge scientists, engineers, and educated lay observers acquire about the fields of genetic engineering and nuclear power, the less fear they have about transgenic crops and nuclear power (Brand 2010: 118).

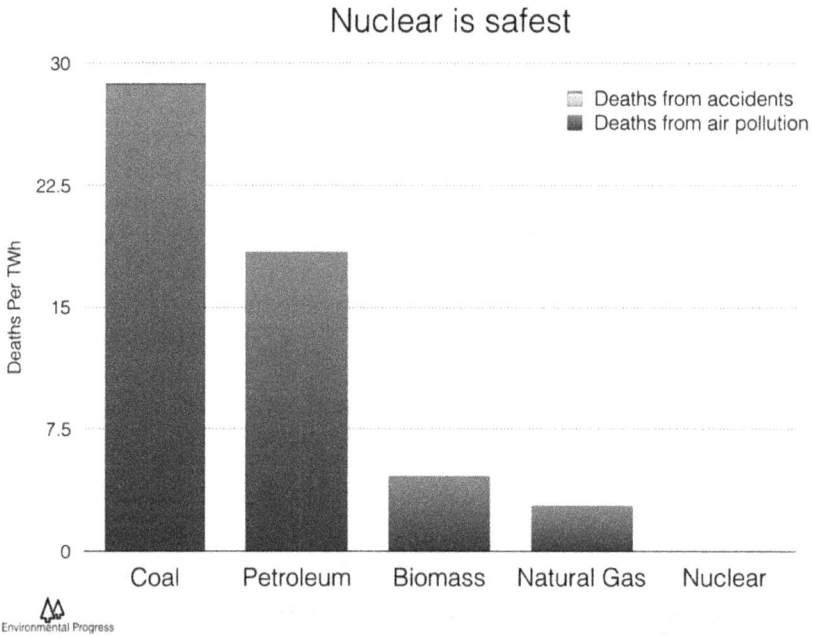

Figure 13.1   Health effects of electricity generation in Europe by primary energy source. *Source*: Courtesy of Environmental Progress.

Nuclear power plant accidents at Three Mile Island and Chernobyl grabbed headlines and released radiation into the environment, yet no human deaths or detectable injuries resulted from the 1979 accident in Pennsylvania; no increases in birth defects among humans or animals have been found in the Chernobyl area since 1986 even after exhaustive medical and biological studies of the region (Brand 92). The biologists Ronald Chesser and Robert Baker have pointed out that the human evacuation from the Rhode Island-sized region around Chernobyl has unintentionally produced one of Europe's most flourishing animal sanctuaries: "The elimination of human activities such as farming, ranching, hunting, and logging are the greatest benefit, and it can be said that the world's worst nuclear power plant disaster is not as destructive to wildlife populations as are normal human activities. Even where the levels of radiation are highest, wildlife abounds" (Brand 94). In a similar way, the evacuation of human residents from the Fukushima area after the tsunami-triggered nuclear accident there was followed by a dramatic influx of wildlife such as wild boars, which have invariably appeared healthy and often even aggressively feisty in spite of being too radioactive to be considered safe for human consumption as wild game (Taylor).

All three abovementioned nuclear accidents occurred at aging first-generation or second-generation nuclear plants that have been increasingly superseded by safer designs such as passively cooled reactors that are not subject to meltdowns or explosions even in the event of an unmonitored plant crippled by a cut-off of electricity (Oberhaus 2020). Furthermore, the processed nuclear fuel that powers most reactors is far too low in density or enrichment (generally containing around 4 percent radioactive isotopes) to be able to be siphoned off for direct employment in nuclear weapons. This is why numerous countries such as Japan, Taiwan, and South Korea have no nuclear weapons programs and yet still have a thriving nuclear energy industry and are signatories of the global nuclear nonproliferation treaty (Li Min 75). Typically, breeder reactors and other special nuclear reactors that produce weapons-grade nuclear fuel in nuclear-armed nations such as the United States and Russia run on a separate track from nuclear reactors that are specifically designed to generate electricity. In the PRC, Mao Era nuclear reactors were nearly all designed to produce highly enriched weapons-grade nuclear fuel, while post-Mao nuclear reactors for electricity generation have been running on a separate lower-enrichment track and design for civilian uses.

Civilian applications of nuclear energy have also saved innumerable lives, especially in high-tech therapies for cancer patients. "Nuclear radiation as used in medicine for diagnosis and treatment has saved countless lives, while exposing all the patients to levels of radiation that are many times what is illegal in the nuclear power industry" (Brand 95).

ELECTRIC GENERATION CAPACITY, RELIABILITY, AND FOOTPRINT OF NUCLEAR VS. ALTERNATIVES

The capacity factor for utility-scale electric generation refers to the percentage of the time during a given year that an electric power plant is producing at its maximum level or full capacity. Nuclear plants far outstrip other forms of electricity generation with their lofty capacity factor of 93 percent—compared to natural gas at 58 percent, coal at 54 percent, hydropower at 43 percent, wind at 37 percent, and solar at 26 percent (US Department of Energy, rounding to the nearest percent).

Nuclear power plants operate at full capacity much more often than other forms of large-scale electricity generation for a number of reasons. For one thing, less maintenance is typically required for nuclear plants than for other types of utility-scale power plants. Soot and other equipment contaminants produced by fossil fuel combustion do not occur as a byproduct of nuclear fission in a reactor. Furthermore, nuclear plants typically operate for longer durations of time prior to refueling—generally 1.5 to two years—than fossil fuel plants, which must be shut down for more frequent refueling. Hydropower, wind, and solar utility plants produce energy merely intermittently due to their need for ample water, wind, or sunlight, the supply of which nearly always fluctuates a great deal throughout a given day or week and from one season to another. Due to the current and projected lack of affordable grid-scale battery storage capability for these intermittent renewable energy sources, engineers must ordinarily rely upon either nuclear or fossil-fuel power plants for the grid's reliable baseload power.

In light of the widely varying capacity factors mentioned above, a one-gigawatt nuclear power plant produces roughly as much electricity as two one-gigawatt coal plants or three or four one-gigawatt wind turbine farms or solar power stations. Nuclear plants thus leave a significantly smaller footprint on the land and exert less pressure on precious wilderness areas or other undeveloped lands than alternative forms of electricity generation. A one-gigawatt nuclear plant occupies no more than a couple of square kilometers of land, while to generate the same amount of electricity, a solar voltaic farm would require over a hundred square kilometers, and a wind turbine farm would be need over four hundred square kilometers (Xu 63; Brand 81). It is little wonder that eco-pragmatists like Stewart Brand who wish to preserve and expand wilderness areas favor nuclear energy in part from a standpoint of wildlands and wildlife conservation.

RELATIVE COMPACTNESS OF NUCLEAR FUEL AND
NUCLEAR WASTES COMPARED WITH COAL

One kilogram of processed uranium-235 reactor fuel with the usual 4 percent concentration of the radioactive isotopes used by power plants produces as much electricity as 1,500 tons of coal (Thorium Power Canada). This giant disparity relates to the much greater energy released in a nuclear fission reaction as opposed to the chemical reaction that occurs when coal, oil, or natural gas is rapidly oxidized during combustion. This huge difference in the amount of fuel needed for electricity generation means that much less damage is done to the land and environment in general when mining uranium than when mining coal. Furthermore, far less energy is required to transport compact rods or pellets of the radioactive fuel to nuclear power plants than to ship hundreds or thousands of tons of coal to coal-burning utility plants. When small module reactors and thorium-based molten salt nuclear reactors increasingly come on line to supersede uranium-based pressurized water reactors (PWR), the impact on the environment of mining nuclear fuel ore will be reduced even further, since thorium is three times more common in Earth's crust than uranium as well as being more evenly dispersed geographically than uranium. Furthermore, thorium may be the only practical type of fuel for large-scale electricity generation that is widely available on the moon (Sauser 2009). Plans are already under consideration at NASA to base a future moon colony's energy supply and life-support systems on a type of thorium-based molten salt nuclear reactor, which unlike standard civilian PWRs do not rely at all upon water for either steam power or cooling (Sauser 2009).

Just as the amount of nuclear fuel needed to power plants is miniscule in comparison to the amount of coal or natural gas required to produce the same amount of electricity, the amount of wastes produced by nuclear plants is also smaller by far. The twenty tons of fuel that a one-gigawatt nuclear plant uses annually are converted into twenty tons of waste that it so dense that it fits into two storage cylinders, each three meters in diameter and six meters tall. These wastes can be safeguarded in dry storage, such as at a large storage site in New Mexico that has been keeping nuclear wastes secure for decades (Brand 81). In contrast, a one-megawatt coal-burning power plant burns three million tons of fuel annually while spewing seven million tons of carbon dioxide into the atmosphere, where it is beyond anyone's control and wreaks havoc with Earth's climate through increasing average temperatures and acidifying the oceans. In addition, coal-fired plants' fly ash and other smoke-stack gases produce more of Earth's human-generated radioactivity than any other source, along with toxic heavy metals such as mercury, arsenic, and lead (Brand 81). As Brand notes, "The air pollution from coal burning is esti-mated to cause 30,000 deaths a year from lung disease in the United States, and 350,000 a year in China" (82). The PRC already had twenty out of the world's thirty most polluted cities in 2000, with coal the main culprit (Ebrey 349). No matter whether a decision maker is looking at the input of fuel or the output of wastes such as coal ash and smokestack gases, coal energy's footprint is vastly larger and deeper than that of nuclear energy.

NUCLEAR'S COSTS: CHEAPER TO RUN OVERALL, BUT LARGER START-UP EXPENSES

In order to make large-scale nuclear power plants reliable and safe with structures such as thick steel reactor vessels and reinforced concrete contain- ment buildings, up-front costs are relatively high. Electric utilities frequently require some assistance from the government in financing plant construction and enhancing insurance coverage once the nuclear plant commences opera- tions, typically a few years after the onset of a plant's construction. Nuclear's relatively high start-up costs are also typical for other types of clean or "green" energy such as wind turbine arrays, hydroelectric dams, and solar photovoltaic farms; solar tends to be the most expensive of all as measured by up-front capital costs per unit of electricity generated.

Yet nuclear energy's cost structure improves a great deal in its operational phase, with average generation costs both lower and more predictable than is typically the case for coal-fired and gas-fired plants (Lillington 30). As Bill Gates has noted, nuclear fission generates about a million times more energy than the combustion of hydrocarbons, so there is far less fuel to mine and transport to a nuclear plant, and longer intervals between shutting the plant for refueling than is the case for plants fired by coal, gas, or oil (Biello). Furthermore, nuclear plants are merely half as sensitive to uranium price volatility as coal-fired plants are to coal price volatility, and merely one-third as sensitive to fuel price volatility as gas-fired plants are to gas price volatility (Lillington 30). The overall result is that the operating costs of nuclear plants are lower than fuel-hungry fossil fuel plants, amounting to merely one-fourth to one-third of a nuclear plant's total expense over its lifetime (Brand 2009: 100).

A STATUS REPORT ON NUCLEAR POWER IN THE PRC AS OF SPRING 2019

Overall, Post-Mao China has made an impressive shift into nuclear power generation, given that it started from essentially zero during the three decades of the Mao Era, when the party-state was overwhelmingly preoccupied with military applications such as stockpiling nuclear bombs and launching nuclear-powered submarines. To be sure, the PRC government appears to have fallen short of achieving its goal to generate fifty-eight gigawatts with its nuclear plants by 2020. Nonetheless, as of spring 2019, the PRC was operating forty-eight nuclear plants that generate a total of approximately forty-three gigawatts, and has under construction another eleven nuclear plants with a total generating capacity of about eleven gigawatts. There are plans for an additional thirty-six gigawatts of capacity going forward, but construction of new nuclear plants tapered off a bit in the PRC during the 2010s (World Nuclear News).

Among the various nations in the world, the PRC has the third largest number of nuclear plants, and the fourth largest amount of nuclear generating capacity. The PRC generates approximately 10 percent of the world's total nuclear energy. However, nuclear power's share as a percentage of the PRC's total electricity generation is less impressive, having increased from 3.9 percent of total electricity produced in 2017 to 4.2 percent of total electricity generated in 2018. This is merely one quarter of what the IPCC recommends as the near-term goal of nuclear power's share in a given nation's overall electricity generation. The China General Nuclear Group (CGN) and other major PRC nuclear power conglomerates may need to pick up the pace a bit if the country is to narrow the gap with IPCC recommendations and achieve the laudable goal of from 120 to 150 gigawatts of nationwide nuclear capacity by 2030 (Stanway).

ATTITUDES TOWARD NUCLEAR POWER PLANTS IN SELECTED PRC SCIENCE FICTION NARRATIVES

Popular attitudes toward nuclear power plants in the PRC may be gauged in part from how home-grown science fiction writers portray this type of electricity generation plant. Overall, a science-based recognition of the practical value and efficiency of nuclear power plants tends to sit uneasily beside a sort of gut-level guilt by association with nuclear bombs and an unfamiliarity with the real (versus imagined) impact of radiation leaks on human health.

Many science fiction writers understand that fossil fuels are unique to Earth within our solar system, thus leaving nuclear power about the only practical alternative energy source for future human colonies on the Moon and Mars. For the human colonists on Mars in Zheng Wenguang's 郑文光 novel *Descendants of Mars* 战神的后裔 rationally generate electricity on Mars through nuclear power plants. This works well for quite a long time before supposedly setting off a giant and destructive nuclear explosion reminiscent of the Hiroshima blast (Zheng 1983). In reality, any explosion from a nuclear power plant meltdown such as Chernobyl is chemical rather nuclear in nature—often from steam or hydrogen gas—and the main danger to human health is the resulting release of radiation from a ruptured containment vessel, not the chemically-based explosion itself.

The science fiction novelist Liu Cixin 刘慈欣makes a similar error in Chapter 22 of his 2005 novel *Ball Lightning* 球状闪电, conjuring forth a scientifically impossible scenario of tiny pill-size uranium pellets that would supposedly trigger a bona-fide nuclear explosion inside a PRC nuclear power plant and thereby cause the death of "hundreds of thousands" of people in the vicinity (Liu 2018: 244, 250). In reality, the uranium fuel used in electricity generation facilities is enriched merely to about 4 percent of fissile material, far below the level required for a nuclear explosion. Moreover, tiny pill-size pellets of even highly enriched uranium are simply too small to achieve the sort of critical mass required to trigger a nuclear fission blast.

On the other hand, Liu Cixin at least presents this nuclear power plant as an efficient, valuable, and ordinarily safe facility prior to its coming under attack by a foreign antitechnology terrorist group called "Garden of Eden" (Liu 2018: 243). Moreover, Liu does not show any sympathy for the anti-nuclear foreign terrorists, whose bloody-mindedness is reminiscent of the US "Unabomber" Ted Kaczynski. In fact, Kaczynski had been arrested less than a decade before Liu's novel was published.

Liu's alter-ego narrator feels sick at heart at the end of the successful defeat of the terrorists not out of sympathy for any of the dead terrorists, but solely because all of the local Chinese schoolchildren and their teacher on a field trip to the nuclear power plant also perished during the security authorities' high-tech military attack on the terrorists inside the power plant. All in all, Liu Cixin's nuclear power plant comes across as not only a major well-func-tioning asset in the PRC's energy portfolio, but also thoroughly defensible in the face of an insidious and well-organized terrorist attack by fanatical foreign luddites. Indeed, the world has yet to experience a single successful terrorist attack on a nuclear power plant that would spread dangerous levels of radiation far and wide such as in a "dirty bomb."

The science fiction writer Yang Dantao 杨丹涛 presents a postapocalyptic nuclear power plant as reliable, clean, and efficient but its chief manager as corrupt and manipulative in his short story "Uranium Flowers" 铀花 (2004). To be sure, science and technology can be utilized for either benevolent or malevolent purposes, depending on the values of the individual or organiza-tion in charge of a given high-tech apparatus. The aspect of Yang's story that strikes me as representing a widely held popular misconception is the reason Yang provides for the local "uranium people's" tolerance for very high levels of radioactivity: they have supposedly inherited genetic mutations due to high levels of radioactivity after a nuclear war many decades ago.

In actuality, while high levels of radiation exposure can indeed result in ill-ness or even death, there is no evidence from the disasters of the explosions at Chernobyl or even Hiroshima that birth defects or other drastic genetic muta-tions that could be passed down to offspring occurred as a result of radiation exposure. Instead, high levels of radiation tend to increase the likelihood of cancer, especially lymphatic cancer. Acute Radiation Syndrome (ARS) can include damage to the patient's DNA, but this affects only the patient and cannot be handed down to the patient's offspring as birth defects in the man-ner that Yang and some other science fiction writers have often assumed and portrayed. Birth defects are far more likely to result from a mother's ingest-ing of dangerous chemicals like thalidomide that have nothing to do with radiation. Instead, severe (and rare) cases of ARS most typically involves one or more of the following afflictions: cancer; burns to the skin; thyroid gland malfunction due to absorption of radioactive iodine; bleeding in the intestinal tract; increased likelihood of seizures; bone marrow and blood ves-sel problems; and inflammation of the lungs. None of the more likely health problems from ARS should be minimized, but neither should writers continue to endlessly spread the popular fallacy that ARS produces birth defects in the next generation. Liu Cixin's *Ball Lightning* repeats the same fallacy: if an explosion were to occur in the nuclear power plant, "the next generation would have deformities" (Liu 2018: 255).

To be sure, Liu has admitted in an interview that both specialist articles in physics and even college-level physics textbooks are typically beyond his ability to comprehend—he is primarily a fiction writer, after all (Nbd.com 2017). Still, PRC science fiction writers have often doubled as popularizers of scientific knowledge, similar to Isaac Asimov in the West. To the extent possible, these science fiction writers would do well to learn more about the genuine evidence-based benefits and risks of nuclear power, including the actual health effects of ARS from the extremely rare cases of nuclear accidents.

CONCLUSION

The urgency of reducing greenhouse gas emissions and slowing global warm-ing requires that influential East Asians and others around the world view the necessary transition to alternatives to fossil fuels with a sober and evidence-based approach to making comparisons. This chapter has noted nuclear power's advantages over its competitors such as its proven health and safety record, its unmatchable level of constant generating capacity, its relatively small footprint on the land, its low fuel costs and modest operating expenses, and its absence of carbon emissions and other air pollutants. While renewables such as solar and wind power certainly deserve continued support and a role in the overall mix of energy sources, their generating capacity at utility scale is severely limited by intermittency and their large footprint on the landscape; in spite of rapid growth during the 2010s, solar power still produces no more than two percent of the US's electricity, merely one-tenth of what the US's nuclear power plants generate (Holthaus). It is thus not surprising that nuclear power has also been an essential tool in the PRC's attempts to lower its dependence on its dirtiest and most dangerous power plants fired by coal (Xu 62). Though radiation accidents are a risk for nuclear power plants, they have been extremely rare, and radiation's health dangers have frequently been exaggerated and misunderstood by popular writers and commentators who are poorly read in evidence-based studies by qualified experts on medicine and science in general.

Nuclear power remains the most viable challenger to fossil fuels in providing base-load electric power twenty-four hours a day and seven days a week, oblivious to wide fluctuations in sunlight, wind patterns, and the water level of reservoirs behind hydropower dams. As the famous atmospheric physicist James Hansen has sagaciously remarked, "Environmentalists need to recognize that attempts to force all-renewable policies on all of the world will only assure that fossil fuels continue to reign for base-load electric power, making it unlikely that abundant affordable power will exist and implausible that fossil fuels will be phased out" (qtd. in Biello).

WORKS CITED

Biello, David. "How Nuclear Power Can Stop Global Warming." *Scientific American*, Dec. 2013. https://www.scientificamerican.com/how-nuclear-power-can-stop-glo bal-warming. Accessed April 20, 2019.

Brand, Stewart. *Whole Earth Discipline: Why Dense Cities, Nuclear Power, Transgenic Crops, Restored Wildlands, and Geoengineering Are Necessary.* Penguin Books, 2009.

Conca, James. "Any New Green Deal Is Dead Without Nuclear Power." *Forbes*, 21 March 2019. https://www.forbes.com/sites/jamesconca/2019/03/21/green-new -deal-is-dead-without-nuclear-power/#68d4aee569db. Accessed July 14, 2020.

Ebrey, Patricia Buckley. *Cambridge Illustrated History of China.* 2nd ed. Cambridge University Press, 2010.

Ellis, James. O, Jr. and George P. Schultz. "The Benefits of Nuclear Power." *Defining Ideas: A Hoover Institution Journal.* https://www.hoover.org/research/benefits -nuclear-power. Accessed April 2, 2019.

Findlay, Trevor. *Nuclear Energy and Global Governance: Ensuring Safety, Security, and Non-proliferation.* Routledge, 2011.

Gibbs, Jeff, director. *Planet of the Humans.* Produced by Michael Moore. Huron Mountain Films, 2020. https://www.youtube.com/watch?v=Zk11vI-7czE. Accessed July 23, 2020.

Harder, Amy. "Environmental Group: Keep Open Nuclear Power Plants." *Axios*, November 8, 2018. https://www.axios.com/environmental-group-keep-open-nuclear-power-plants-6177b5b0-3885-48f3-a51a-4a76436609a3.html . Accessed April 13, 2019.

Holthaus, Eric. "It's Time to Go Nuclear in the Fight Against Climate Change." *Grist*, 12 Jan. 2018. https://grist.org/article/its-time-to-go-nuclear-in-the-fight-against-climate-change/. Accessed March 30, 2019.

Li Min 李敏. "Taiwan bu neng fangqi heneng" 台灣不能放棄核能 [Taiwan Cannot Abandon Nuclear Energy]. *Kexueren* 科學人 [Scientist], no. 6, 2011, pp. 74–75.

Lillington, John. *The Future of Nuclear Power*. Elsevier, 2004.

Liu, Cixin 刘慈欣. "Di ershi'er zhang" 第二十二章 [Chapter 22], *Qiuzhuang shandian* 球状闪电 [Ball Lightning]. Sichuan kexue jishu chubanshe 四川科学技术出版社, 2005.

———. *Ball Lightning*. Translated by Joel Martinson. Tom Doherty Associates, 2018.

Markandya, Anil and Paul Wilkinson. "Electricity Generation and Health." *Lancet*, vol. 370, no. 9591, September 15, 2007, pp. 979–990. DOI: https://doi.org/10.1016/S0140-6736(07)61253-7

Nbd.com. "Zhuanfang kehuan zuojia Liu Cixin: 'Wo he nimen yiyang kanbudong qianyan wuli lunwen'" 专访科幻作家刘慈欣: "我和你们一样看不懂前沿物理论文"[Special interview of science fiction novelist Liu Cixin: "Like all of you, I can't understand cutting-edge physics articles"]. 每经网 [Nbd.com], December 19, 2017. http://www.nbd.com.cn/articles/2017-12-19/1173292.html. Accessed July 19, 2020.

Oberhaus, Daniel. "Nuclear 'Power Balls' May Make Meltdowns a Thing of the Past: This Alien-Looking Fuel with Built-in Safety Features Will Power a New Generation of High-Temperature Reactors." *Mother Jones*, July 6, 2020. https://www.motherjones.com/environment/2020/07/nuclear-power-balls-may-make-meltdowns-a-thing-of-the-past/. Accessed July 15, 2020.

Office of Nuclear Energy, U.S. Department of Energy. "Nuclear Power Is the Most Reliable Energy Source and It's Not Even Close." https://www.energy.gov/ne/articles/nuclear-power-most-reliable-energy-source-and-its-not-even-close. Accessed April 29, 2019.

Rhodes, Richard. Review of Jared Diamond's *Upheaval: Turning Points for Nations in Crisis*. In *Nature*, vol. 568, April 18, 2019, pp. 312–314.

———. "Why Nuclear Power Must Be Part of the Energy Solution." *Yale Environment 360*, July 19, 2018. https://e360.yale.edu/features/why-nuclear-power-must-be-part-of-the-energy-solution-environmentalists-climate. Accessed March 28, 2019.

Sauser, Brittany. "A Lunar Nuclear Reactor: Tests Prove the Feasibility of Using Nuclear Reactors to Provide Electricity on the Moon and Mars." *M.I.T. Technology Review*, August 17, 2009. https://www.technologyreview.com/2009/08/17/210973/a-lunar-nuclear-reactor/. Accessed July 23, 2020.

Stanway, David. "China's Total Nuclear Capacity Seen at 120-150GW by 2030—CGN." *Reuters*, March 14, 2016. https://af.reuters.com/article/commoditiesNew

s/idAFL3N16M3QXhttps://af.reuters.com/article/commoditiesNews/idAFL3N1
6M3QX . Accessed April 17, 2019.

Taylor, Alan. "The Wild Boars of Fukushima." *The Atlantic*, March 2017. https://ww
w.theatlantic.com.photo/2017/03/the-wild-boars-of-fukushima/519066/ . Accessed
April 19, 2019.

Thorium Power Canada, Inc. "Thorium vs. Uranium Fuels." 2015. http://www.thor
iumpowercanada.com/technology/the-fuel/thorium-vs-uranium-fuels/. Accessed
April 25, 2019.

Union of Concerned Scientists. "Nuclear Power and Global Warming." https://www
.ucsusa.org/nuclear-power-and-global-warming. Accessed April 18, 2019.

World Nuclear Association. "World Energy Needs and Nuclear Power." Updated
February 2019. www.world-nuclear.org/information-library/current-and-future
-generation/world-energy-needs-and-nuclear-power.aspx. Accessed March 31,
2019.

World Nuclear News. "Start-up Nearing for Chinese Units." *World Nuclear News*,
March 25, 2014. http://www.world-nuclear-news.org/NN-Start-up-nearing-for-Ch
inese-units-2503144.html. Accessed April 22, 2019.

Xu Yi-chong. *The Politics of Nuclear Energy in China*. Palgrave Macmillan, 2010.

Yang Dantao 杨丹涛. "You hua" 铀花 [Uranium Flowers]. *Kehuan shijie* 科幻世界
[Science Fiction World], vol. 216, no. 5, 2004, pp. 30–37.

Yi Si. "The World's Most Catastrophic Dam Failures: The August 1975 Collapse of
the Banqiao and Shimantan Dams." In *The River Dragon Has Come! The Three
Gorges Dam and the Fate of the Yangtze River and Its People*. By Dai Qing; edited
by John J. Thibodeau and Philip B. Williams. M.E. Sharpe, 1998, pp. 25–38.

Zheng Wenguang 郑文光. *Zhanshen de houyi* 战神的后裔 [Descendants of Mars].
Huacheng chubanshe 花城出版社, 1983.

Chapter 14

The Revitalization of Old Industrial Sites in Beijing

A Case Study of *Shougang* (Capital Steel) Park

Xin Ning

INTRODUCTION

This chapter aims to examine the revitalization process of old industrial sites in Beijing, the capital of the People's Republic of China (PRC), with focus on one particular case: the transformation of the site of *Shougang* (Capital Steel) Company. Occupying an area of 8.63 km², the *Shougang* site, which is located in *Shijingshan* District, northwest of Beijing, is the largest one among the city's old industrial sites under redevelopment. The transformation of *Shougang* site started at the beginning of the new millennium: To decrease the pollution that iron and steel production brought to the city to make preparation for 2008 Summer Olympics, *Shougang* Company gradually relocated its manufacturing branches outside Beijing since 2002 and finally stopped its iron and steel production in Beijing in 2011. The original site was redesigned into "The New *Shougang* High-end Industry Comprehensive Service Park" (Li, "Shougang lishi yange" 16–17; "Shougang Park"). Currently, the *Shougang* Park hosts The Organizing Committee of the XXIV Olympic Winter Games, Innovation Center for Overseas Chinese Businesspersons, China's first and the world's 19th C40 Climate Positive Development Program project, and National Sports Industry Demonstration Zone. During the Winter Olympics to be held in 2022, the *Shougang* Park will serve as a major training center for short track speed skating, figure skating, curling, and ice hockey. A big single-board skiing jump is also to be built here ("Aoyun shinian Shougang zhi bian"). The overall development plan claims that the *Shougang* Park will make itself "a demonstration zone of green transformation of the traditional industry and a . . . post-industrial cultural and sports creative base in western Beijing" and plans to "build three leading industries in sports+, digital intelligence and cultural creativity" ("Shougang Park"). My study of the massive transformation of *Shougang* site will address a series of questions in seemingly diversified fields which, in the final analysis, prove to be closely interrelated: the rethinking of the connections of city dwellers and their lifeworld in the light of urban ecology; the remapping of urban landscape and the development of creative economy in the postindustrial city, and the entanglement between Neo-Bohemian community of artists, commerce, and government in city redevelopment projects. The study of the *Shougang* case would illustrate China's specific responses to such questions on the one hand, and contribute to a broad understanding of the relationship between ecology, creative industry, and urban development in the age of globalization on the other.

FROM MODEL FACTORY TO OLYMPIC PARK

The beginning of *Shougang* Company can be traced back to the foundation of Shijingshan *lianchang* (Shijingshan Smelting Factory) in 1919 (*Beijing shi Shijingshan qu zhi* 528). We may argue that the history of *Shougang* Company epitomizes the general history of the development of modern industry in China. The first group of modern enterprises in China emerged in mid-nineteenth century after the door of China was forced open in the First and Second Opium Wars (1839–1842; 1856–1860). In the 1860s, modern manufacturing factories began to be established in China, first by foreign industrialists and then by Chinese government and native bureaucrats and merchants during the self-strengthening movement (*yangwu yundong*). Although Beijing, as the capital of the old empire, was more hesitant than coastal cities such as Shanghai and Tianjin to embrace the new tide of industrialization, we find on record that the first modern industrial enterprise in Beijing, *Tongxing* Coal Mine, was founded as early as 1872 (Cao 8: 296). The modern coal industry in the northwestern suburb of Beijing later became one of the decisive elements for the choice of the location of Shijingshan *lianchang*. The smelting factory was a joint-venture enterprise with both state and private investments until being nationalized in 1928. Yet the building of the factory was stopped in 1922 and not resumed until 1938. Ironically, it was during the Japanese occupation period (1937–1945) that the construction of the factory was finally finished and the production of iron (but not steel) began. Until the end of 1948, the factory had only produced 286,200 tons of iron *in toto*. After the founding of the People's Republic of China in 1949, the company underwent quick resurrection and soon became a star enterprise in the new Socialist Era. In 1957, it began to produce steel for the first time. In 1994, *Shougang Steel* became the largest steel producer in China (*Beijing shi Shijingshan qu zhi* 528–530; Li, "Shougang lishi yange" 13–16).

Shougang Steel had played the role of a model enterprise in both the Socialist and Early Reform and Opening-up era of China by successfully meeting the expectations of the growth-oriented and technicist understanding of progress that so far dominated Chinese society. Such understanding puts priority on material production and economic growth, often at the expense of nature and to the neglect of human needs. The performance of an individual enterprise would be judged either by the sheer amount of its output, as in the Socialist Era, or the net profit and workforce productivity, as in the new era of Reform and Opening-up. As the capital of the new People's Republic, Beijing in particular was designed to become an industrial center (which it had never been in history) and *Shougang Steel* was a jewel on the crown. Only with the awakening of modern ecological consciousness and the rethinking of the pattern of development in 1990s then did people begin to give serious thought to the environmental, social harms and even crises, that the iron and steel industry in the metropolitan area may have led to. The shadows beneath the brilliant record of *Shougang Steel* were gradually brought to the attention of the public and eventually affected government policy making.

From the environmental point of view, the decision of establishing an iron and steel factory in the northwest of Beijing is a hazardous mistake from the beginning. Beijing is surrounded by maintains in its north, west, and east and open to plains in its southeast. The major rivers of Beijing all originate from the northwestern mountains and flow through the city to the southeast plains. Also in winter time, harsh northwest wind blows from the mountainous area and sweeps through the entire city (Hou and Deng 6–8). *Shougang Steel*, nonetheless, is located right on the northwestern highland. Such a choice severely aggravates the pollution that is inevitable in the iron and steel industry. The founders of *Shougang* Company one century ago obviously gave no consideration to environmental issues and chose the site merely for the convenience of fuel and raw material supply. Moreover, with the quick expansion of Beijing in the latter half of the twentieth century, *Shougang Steel* is eventually incorporated into the urban area of the metropolis and embedded within a densely populated residential district. The air, water, and soil pollution of *Shougang Steel* finally became unbearable. For instances, investigations in the 1980s and 1990s showed that the water body of Xinkai Inlet Channel of Yongding River in the area was "heavily polluted" and the amount of suspended solids and sulfur dioxide in the air of Shijingshan District was constantly above the national standard, sometimes doubling it (*Beijing shi Shijingshan qu zhi* 397–404). *Shougang Steel* is a major contributor to both the water and air pollution. In the 1990s, the iron and steel production of *Shougang Steel* brought thirty-four tons of dust per square kilometer per year to Shijingshan District (Li, "Shougang xiemu Beijing" 5). Also large amounts of iron, cadmium, phenol, and other metals and chemicals have poisoned the soil and the groundwater on the site of *Shougang Steel* (Wang and Wang 212). Indeed, some experts have initiated a debate on "either to move *Shougang* or to move the capital" as early as in the 1980s. The debate reemerged in the late 1990s, partly due to the coming of 2008 Summer Olympics in Beijing, and finally resulted in *Shougang Steel*'s decision of overall relocation and reorganization, which was officially approved by the State Council in 2005 (Li, "Shougang xiemu Beijing" 5). Yet it takes another twelve years for the redevelopment plan of the original *Shougang Steel* site in Beijing to take shape. With lots of delays and vicissitudes, the latest version was finalized and officially approved in 2017. The plan sets the entire Beijing *Shougang Steel* site as The New *Shougang* High-end Industry Comprehensive Service Park and divides it into the North District and South District. The North District includes Winter Olympic Game Square, *Shougang* Industrial Relics Park, Shijingshan Sightseeing Park, "Weaving" City Innovation Workshop and Public Service Section. The detailed plan for the South District has not been decided yet but the district will probably be devoted to the service for overseas Chinese businesspersons ("Jiangxin dazao chengshi fuxing xin dibiao").

The *Shougang* Park in current design is the agglomeration of various subareas and functions, whose successful integration seems rather problematic. Once finished, the park will simultaneously be a center for winter sports, a central business district (CBD) with luxurious offices and apartment buildings, an incubator and testing ground for start-up companies in creative economy, a memorial site for early industrialization in the Socialist Age, and the urban greenspace with plantation, grassland, gardens, and water features. It is also noticeable that the governments at all levels—state, municipal and local—play a dominant role in the redevelopment project: The plan is officially drawn and approved by the government and is to be implemented by the government largely with government funding. Government officials routinely visit *Shougang* Park and give directions on the redevelopment project, and the state media enthusiastically makes detailed reports on their activities. For example, the Secretary of Beijing Communist Party Committee, Cai Qi, and the Mayor of Beijing, Chen Jining, recently took a tour of inspection around *Shougang* Park on August 25, 2018, urging the realization of "city resurrection" through the project (Qi and Wu). Local TV channels and newspapers soon made full coverage on their visit and talks in the following two days. Like the *Shougang Steel* in the old days, the current *Shougang* Park is still expected to become a model in the eye of the central government.

These two features of the *Shougang* Park—its complicated and some-how self-contradictory orientations and the government dominance in the project—distinguish it from most of other revitalization projects in Beijing. Indeed, we may argue that the *Shougang* Park project represents a new pattern for city revitalization that the government is eager to implement in the future. Whether it is more productive, sustainable and ecofriendly than the patterns already adopted, however, is an open question.

CREATIVE ECONOMIES AND URBAN ECOLOGY ON THE OLD INDUSTRIAL SITES

The decline of manufacturing industry and the burgeoning of knowledge or creative economy in recent years have left their marks on the urban landscape of numerous cities around the world. Lots of old industrial bases, besides being leveled down to make space for new office towers and luxurious apartment buildings, are transformed into artists' workshops, cultural centers, and theme parks. The nature and impact of such a "cultural turn" of former industrial sites, however, turn out to be ambiguous and provoke heated debates among intellectuals, artists and official policy-makers.

Richard Floria optimistically claims in *The Rise of the Creative Class* that we are living in the age of creative economy and witnessing the emergence of a new creative class. He emphasizes the key role of innovation in today's economy:

> I see creativity—the faculty that enables us to derive useful new forms from knowledge—as the key driver of today's economy. In my formulation, knowledge and information are merely the tools and the materials of creativity. Innovation, whether in the form of a new technological artifact or a new business model or method, is its product (30).

The persons who engage in creative economy form the creative class, of whom "if you are a scientist or engineer, an architect or designer, a writer, artist, or musician, or if your creativity is a key factor in your work in business, education, health care, law, or some other profession, you are a member" (xxi). The rise of creative economy and the new, self-conscious creative class requires the remapping of unban space. According to Floria, creative economy insists on the principles of both cultural diversity and ecological livability and sustainability and encourages modern urban communities' attempts of "reclaiming their disused waterfronts and industrial areas and transforming them into parks and green spaces . . . [to] make people's everyday lives better, improve the underlying quality of place, and signal a community that is open, energized, and diverse" (x).

The burgeoning of art zones on old industrial sites in Beijing in the past two decades seem to echo Floria's call for the transformation of urban space to meet the need of new economy and new class. So far dozens of such "art zones" have been developed in the city, attracting independent artists all around China to set their studios, exhibit their works and at the same time form loose yet intimate "Bohemian" communities there. Some of them, such as 798 Art Zone (receiving its name since it is located on the site of decommissioned 798 Factory), have successfully established themselves as new Meccas for avant-garde art and alternative life styles and enjoyed international fame. Further examination of such new urban art zones leaves us suspicious, however, about their role in creating an ecofriendly and human-friendly lifeworld in the postindustrial city. Two questions, one specific and the other more general, have to be asked: Is *Shougang* Park a "typical" postindustrial artist enclave like others in Beijing? Does the replacement of Fordist factories (assuming *Shougang* was one such) with artist enclaves really fit the trend of creative economy and help to create "open, energized, and diverse" communal space in the postindustrial city?

To the first question, we can make a somewhat straightforward answer: *Shougang* Park differs from other revitalization projects in Beijing in both its planning and management strategy and cannot be viewed as an "exemplary" postindustrial artist enclave. I hasten to add that each individual case of renovation and revitalization of old industrial sites is unique and conditioned by its immediate economic, political, social, and cultural context; hence, the idea of an "exemplary" pattern is but a limited if not totally misleading concept. However, at least we can point out that most of other revitalization projects in Beijing started as a spontaneous process initiated by the artist community with minimum governmental interference. The lack of governmental control helps to leave space vacant for individual freedom and stimulate creativity of the artists. The development of *Shougang* Park, on the other hand, is under strict governmental surveillance from the beginning. Instead of a haven for Bohemian artists, it is a model program designed to demonstrate governmental competence in urban management as well as to pursue net profit in the spirit of capitalist entrepreneurship. It is also a big theme park for tourist attraction, whose selling points are nostalgic memory of socialist industrialization on the one hand, and nationalist pride kindled by Olympic games on the other. Without denying its positive role in urban redevelopment and environmental improvement, *Shougang* Park is nonetheless part of what Floria describes as "organization" that is marked by "the dominance of large scale and highly specialized bureaucracies" and in constant tension with creativity yet so far also an indispensable of social machine (16). Indeed, the necessary changes to urban space that Floria has in mind is not the massive renovation project with "stadium complexes and generic retail districts and malls," but down to earth improvement of neighborhood conditions "with smaller investments in everything from parks and bike paths to street-level culture" (x). He agrees with Jane Jacob in the defining features of ideal urban space in the postindustrial and postmodern city. Jane Jacob exemplifies such urban space in her analysis of neighborhoods in Greenwich Village in New York City in her book *The Death and Life of Great American Cities* published in 1961. According to her, the Greenwich neighborhoods compose a communal space where "[p]eople of all classes and educations, with all kinds of ideas, were constantly jostling against each other and striking intellectual sparks" (Floria 28). The cultural diversity and creativity is guaranteed by the physical environment of the neighborhood:

> It had short blocks that generated the greatest variety in foot traffic . . . It had broad sidewalks and a tremendous variety of types of buildings—apartments, stores, even small factories—which meant that there were always different kinds of people outside and on different schedules. There were lots of old, underutilized buildings, ideal for individualistic and creative enterprises ranging from artists' studios to small entrepreneurial businesses (Floria 28).

A quick comparison of *Shougang* Park with Jacob's Greenwich neighborhoods clearly demonstrates their polarized differences and disqualifies the former as the ideal space for either Bohemian communities or the creative economy that Floria keenly hails.

Beijing Municipal Government, however, seems rather proud of the pattern of *Shougang* Park redevelopment and determined to consolidate the governmental presence in all industrial site revitalization projects in the future. On December 31, 2017, it issued *Guanyu baohu liyong laojiu changfang tuozhan wenhua kongjian de zhidao yijian* (Guiding Opinions on the Protection and Employment of Old Industrial Buildings to Expand Cultural Space), which proclaims that all the revitalization projects "must be guided by the government and run by the market." The government would "thoroughly undertake its duties to . . . advance the protection and employment of old industrial buildings" and the related enterprises would "play the central role" in the renovation projects through "marketing process such as rental, assets reorganization, and trusteeship." The industrial enterprise that thinks of engaging in cultural industry on its original site must compose a detailed "Plan for the Protection and Comprehensive Employment" for the individual project and submit its plan for official approval before taking actions to renovate the old buildings and run the art zone. The only positive point in this guide, to me, is that it finally acknowledges the cultural importance of old industrial sites and manages to maintain the buildings instead of pulling them down to make space for new skyscrapers and shopping malls. However, it is disappointing that a document on the "expansion of cultural space" simply emphasizes the dominant role of government in cultural activities and makes strict regulations on the running of the art zone project yet does not mention the actual needs of artist community at all. Neither can artists own their studios in the art zones (the land and buildings are still owned by the State), nor do their opinions really count in the planning and running of the art zones. I am not totally convinced that such pattern of revitalization of old industrial sites, which *Shougang* Park faithfully follows, can successfully stimulate creativity instead of stifling it.

We may now turn to the more general question: does the "cultural turn" of urban industrial sites neatly fit the vision of creative economy and new urban space? At the first sight, the rise of art zones provides us with a very promising picture: artists working as members of creative class, their artistic creation as a coherent part of creative economy, and art zones as components of ideal new urban space. Yet further analysis gives us a more disturbing version of the said story. Even the most triumphant and thriving art zones are far from being Eleusinian havens for artists; instead, each of them takes shape through the collaboration as well as contestation between different interest groups such as the State and City authorities, developers, artists, and local inhabitants. Richard Lloyd, in his study of similar artistic enclaves in American cities, points out:

> The "revitalization" of core city districts through the leveraging of artistic energies produces a mixed bounty, fruitfully countering the low-road, tourism-directed policies that had dominated revitalization strategies in an earlier period, and enriching the cultural landscape, but also abetting the neoliberal tendencies toward cutthroat interurban competition and the promotion of gentrification. Moreover, artists, far from being liberated by the heightened attention to local culture, find themselves co-opted into new forms of postindustrial economic exploitation (Lloyd xii–xiii).

These new art zones, despite their carefree Bohemian atmosphere, are entangled with the neoliberal trend of the commercialization of culture. Firstly, they are under the constant threat of gentrification and various forms of commercial redevelopment plans, initiated either by the enterprises or the local government. As some scholars sarcastically put it, "artists are often the first to lease studios in redundant industrial buildings, but also the first to go when designation as a cultural quarter leads to higher rents" (Bennett and Butler, 59). Indeed, some art zones have been so manifestly commercialized that they began to be labeled Chinese BoBo—Bourgeoisie and Bohemian—space: they contain far more fancy restaurants, bars, cafés, clubs, and luxury shops than artists' studios.

Second, what is more disturbing is that the artists in the art zones are often lured by monetary gains and worldly fame and choose to comply with the taste of potential customers and follow the "fashion." We have to admit that the "independent" artists working in the art zones are still part of the "organization" of domestic and global art market and under direct or indirect control of agents, art dealers, critics, curators and patrons. It is not coincidental, for example, that although there are old industrial sites scattered all around the city, most of postindustrial art zones in Beijing are located in Chaoyang District, where locates Beijing CBD and most of foreign embassies. This is the area with the densest population of foreigners and *nouveau riche* Chinese, including successful businesspersons and high-ranking professionals working in foreign enterprises and international organizations, who are the target buyers of the avant-garde style works that fill the studios in the art zones. The customer is God.

Third, the attempts of changing industrial sites to art zones always involve with complicated processes of deterritorialization and reterritorialization of a large number of people, not only artists but, more importantly, local inhabitants. If the relocation of the former follows a centripetal pattern as the art zones keep drawing artists from various parts of the country to the metropolitan center, then the resetting of the latter group often takes a centrifugal pattern when the original residents are disassembled and dispersed from their old neighborhoods to distant suburban areas of the city. The impact of the transformation of old industrial sites on local inhabitants often does not receive the attention it deserves, however, in the discussions on the revitalization of old industrial sites in the city. Yet there are still critics who perceptively recognize the threats to the "authenticity" of the city brought by the tide of "cultural turn" in urban redevelopment as well as its damaging impact on the local community. Sharon Zukin, for example, is one of the pioneering scholars who sharply points out the capitalist logic behind the "cultural boom" of postindustrial cities:

> With the disappearance of local manufacturing industries and periodic crises in government and finance, culture is more and more the business of cities—the basis of their tourist attractions and their unique, competitive edge. The growth of cultural consumption . . . and the industries that cater to it fuels the city's symbolic economy, its visible ability to produce both symbols and space (*The Culture of Cities* 1–2).

Later he recapitulates his argument with a more detailed case study of New York City in *Naked City: The Death and Life of Authentic Urban Spaces*. As he argues, beginning in the 1970s "developers of downtown shopping centers turned derelict industrial and waterfront land into profitable attractions" and in the 1980s "cultural districts, ethnic tourist zones, and artists' lofts presented a clean image of diversity for mass consumption" in New York (5). However, during this seemingly successful transformation of decayed neighborhoods the "Origins" of the city were irretrievably lost. To Zukin,

> "Origins" suggests . . . a moral right to the city that enables people to put down roots. This is the right to inhabit a space, not just to consume it as an experience. Authenticity in this sense is not a stage set of historic buildings as in SoHo or a performance of bright lights as at Times Square; it's a continuous process of living and working, a gradual buildup of everyday experience, the expectation that neighbors and buildings that are here today will be here tomorrow. A city loses its soul when this continuity is broken (*Naked City* 5–6).

The neighborhoods of the working class in New York City, which Jane Jacob has passionately defended in *The Death and Life of Great American Cities* in the early 1960s, were lamentably displaced through city gentrification, of which the "cultural boom" is a particular type:

> Life in the original "urban village" of ethnic and working-class neighbor-hoods before the 1960s was a re-creation of tradition. In the gentrified and hipster neighborhoods that have become models of urban experience since then, authenticity is a consciously chosen lifestyle and a performance, and a means of displacement as well (*Naked City* 4).

Although Zukin's work concentrates on the study of New York City, he also notices that "Beijing, Shanghai, and other Chinese cities are clearing out the narrow, rundown alleys in their center, removing longtime residents to the distant edges of town, and replacing small, old houses with expensive apart-ments and new skyscrapers of spectacular design," which, like the changes of New York, belongs to a "[a] universal rhetoric of upscale growth, based on both the economic power of capital and the state and the cultural power of the media and consumer tastes" (*Naked City* 1–2). The transformation of old industrial sites to art zones dissolves the local communities and leads to the deterritorialization of original inhabitants, most of whom are members of the city working class. Zukin's description of the decline of working-class neighborhoods in New York City ironically fits the cases of Chinese cities in a more dramatic way. In the Socialist Era before Reform and Opening-up period, the urban society in China was organized around various *danwei*, which is simultaneously a working and living unit. A *danwei* can be a factory, a government department, a school, a hospital, etc. The common characteris-tic of *danwei is* that it is not only a working place but also provides residential buildings and basic social services, from dining halls and barber shops to

primary and middle schools and hospitals, to its employees and their families. Hence, around each old industrial site in the city there is always a living
community composed of workers and their families who have lived there for
decades and even generations. Most of these inhabitants, however, are to be
relocated in the redevelopment plan while the old industrial site is cleaned,
both physically and functionally, for its new role of avant-garde art zone,
theme park, or city museum. The vigorous grow of the "symbolic economy"
of culture is at the price of the dissolution and dislocation of authentic local
community and original inhabitants.

This is the case of both *Shougang* Park and other art zones on old industrial sites in Beijing—ironically, it is perhaps the most remarkable, if not the
only, similarity between them. In the case of *Shougang* Park, between 2005
and 2007, 7,000 workers of *Shougang* Company have been moved out of
Beijing in the relocation process of the company's iron and steel manufacturing branch (Li, "Shougang xiemu Beijing"). It no doubt brings significant
changes to the neighborhoods around *Shougang* Park in Beijing, yet probably not as drastic as we imagine. For most of the workers relocated, their
families still remain in Beijing in their old residential places, and the workers
themselves either commute between Beijing and their new working place or
constantly return during weekends and holidays. In short, there is still a large
working-class community around *Shougang* Park now and the residents of
the community will be inevitably affected by the redevelopment plan in the
coming years. There is no mentioning of the fate of the local residents in the
officially approved redevelopment plan of *Shougang* site, however, and I
wonder if the workers and their families can benefit from the construction of
the "High-end Industry Comprehensive Service Park" for "leading industries
in sports, digital intelligence and cultural creativity" in their familiar living
environment.

The revitalization of old industrial sites is not only a social but also an
ecological issue. Indeed, the social and the ecological concerns are always
closely intertwined with each other: any discussion of social justice must
include ecological justice while effective maintenance of a healthy ecosystem
must take into consideration the human factors and incorporate the ecosystem
into the "totality" of life-world. Here I understand the term totality as Timothy
Morton interprets it: "total interconnectedness" without being necessarily
single, closed, independent, fixed, or predetermined (Morton 40). Indeed, in
the above discussions on Jacob and Floria we have already touched the issue
of urban ecology, for Floria makes it clear that the ideal urban space should
generate and maintain both cultural diversity and ecological livability and
sustainability. Urban ecology, as Ian Douglass defines it, aims to "understand
the full complexity of the relationship between the biological community and
the urban environment due to the interaction between human culture and the

natural environment" (Douglass et al. 3). It is closely related to human ecology that "emphasizes the contribution of the individual, the sharing of skills and experiences, and the dignity and insight of social and cultural and religious experiences . . . [and] works to create sustainable, lasting improvements in people's lives by fostering projects that engage and enhance the skills of local communities, involve all sectors of society, improve livelihoods and maintain environmental benefits" (Douglass et al. 4). As I understand it, human ecology should not be confused with traditional anthropocentrism that regards human beings as the central entities in the world and takes human dominance on nature and other species for granted. Rather, human ecology urges human beings to self-consciously participate in the creation and preservation of a vibrant and healthy ecosystem with a full recognition of the social and ideological factors in such historical practice. Methodologically, though, it may tend to focus on the study of human elements in the ecological and social system as well as the reciprocal relationship between human beings and their environment. As cities and towns are "human-made ecosystems that result from the interrelations between ecological, economic, material, political and social factors" (Douglass et al. 38), however, such methodological choice is a theoretical and practical necessity.

Shougang Park is a controversial case in the light of urban ecology. On the one hand, there is no doubt that the removal of iron and steel industry and the expansion of greenspace benefit the city's environment. On the other hand, the severe soil pollution on the site of *Shougang* Park is not fully addressed in the current redevelopment plan, which leads to concerns about the ecological safety and sustainability of the project. Moreover, the immense redevelopment project marks another step in the seemingly infinite expansion of metropolitan Beijing, an urban development pattern that is unbearable to the city's frail ecosystem, overladen infrastructure, and general social framework. Beijing government has somehow realized the potential dangers of such endless expansion yet found no better solutions than canceling the so-called "noncapital functions" (*fei shoudu gongneng*) and cutting the city's large mobile population. In practice, however, such measures afflict the most fragile community of the city, namely the underprivileged migrant workers and their families, and undermine the basic principles of both social justice and market economy. With its Olympic Park, new CBD district, high-tech laboratories and enterprises and green space, *Shougang* Park fulfills all the expectations for the ideal capital in the blueprint made by the government. Yet besides such well-regulated space for business and elites, a truly livable city still needs some low-key public and communal space for authentic human contacts and class, ethnic and cultural diversity, some cozy messiness, and a sustainable ecosystem that balances environmental protection and human needs. I would not call

Shougang Park or other art zones projects failures, yet the revitalization of old urban industrial sites as well as the overall "city resurrection," a favorite term of Beijing governmental officials, demand sincere exploration of alternative possibilities beyond neo-liberalist policies and designs of development.

WORKS CITED

"Aoyun shinian Shougang zhi bian." 奥运十年首钢之变 (The Changes of *Shougang* During the 2008–2018 Decade). Shougang Group. August 8, 2018. <www.shouga ng.com.cn/sgweb/html/sgyw/20180808/2254.html>. Web. Accessed August 10, 2018.

———. "Jiangxin dazao chengshi fuxing xin dibiao." 匠心打造城市复兴新地标 (Forging the New Landmark of City Renovation with Artisan Spirit). Shougang Group. Jan. 12, 2018. <www.shougang.com.cn/sgweb/html/sgyw/20180112/1646. html>. Web. Accessed August 10, 2018.

———. "Shougang Park." Shougang Group. n.d. <www.shougang.com.cn/en/ehtml/ ShougangPark/>. Web. Accessed August 10, 2018.

Beijing shi zhengfu 北京市政府 (Beijing Municipal Government). *Guanyu baohu liyong laojiu changfang tuozhan wenhua kongjian de zhidao yijian*关于保护利用老旧厂房拓展文化空间的指导意见 (Guiding Opinions on the Protection and Employment of Old Industrial Buildings to Expanse Cultural Space). Beijing Municipal Government. Dec. 31, 2017. <zhengce.beijing.gov.cn/ zfwj/5111/5141/1344481/1426940/index.html>. Web. Accessed August 10, 2018.

Bennett, Sarah and John Butler, eds. *Advances in Art & Urban Futures,* Vol. 1: *Locality, Regeneration & Divers[c]ities.* Bristol: Intellect Books, 2000. Print.

Cao Zixi 曹子西. *Beijing tongshi* 北京通史 (The General History of Beijing), 10 Vols. Beijing: Zhongguo shudian, 1994. Print.

Douglas, Ian et al., eds. *The Routledge Handbook of Urban Ecology.* New York: Routledge, 2011. Print.

Floria, Richard. *The Rise of the Creative Class, Revisited* (10th Anniversary Edition). New York: Basic Books, 2012. Print.

Hou Renzhi and Deng Hui 侯仁之、邓辉. *Beijing cheng de qiyuan yu bianqian* 北京城的起源与变迁 (The Origin and Transformation of Beijing City). Beijing: Chongguo shudian, 2001. Print.

Jacobs, Jane. *The Death and Life of Great American Cities.* New York: Vintage Books, 1961. Print.

Li Shu 李舒. "Shougang xiemu Beijing" 首钢谢幕北京 (The Capital Steel Company's Farewell to Beijing). *Beijing jishi* 北京纪事 2007 (8): 4–12. Print.

———. "Shougang lishi yange." 首钢历史沿革 (The History of the Capital Steel Company). *Beijing jishi* 北京纪事 2007 (8): 13–17. Print.

Lloyd, Richard. *Neo-Bohemia: Art and Commerce in the Post-Industrial City* (2nd ed.). New York: Routledge, 2010. Print.

Morton, Timothy. *The Ecological Thought*. Cambridge, MA: Harvard University Press, 2010. Print.

Qi Mengzhu and Wu Hongli 祁梦竹、武红利. "Dazao xin shidai shoudu chengshi fuxing xin dibiao." 打造新时代首都城市复兴新地标 (Forging the New Landmark of City Renovation In the New Age). *Beijing ribao* (Beijing Daily). August 27, 2018. <http://bjrb.bjd.com.cn/html/2018-08/27/content_276044.htm>. Web. Accessed August 31, 2018.

The Editing Committee. *Beijing shi Shijingshan qu zhi* 北京市石景山区志 (The History of Shijingshan District of Beijing). Beijing: Beijing chubanshe, 2005. Print.

Wang Chaoyi and Wang Zuo 王超益、王佐. "Shougang jiu gongye jianzhu lüse gaizao zailiyong yanjiu." 首钢旧工业建筑绿色改造再利用研究 (The Research about Green Renewal and Reutilization of Old Industrial Architectures of Shougang). *Jianzhu yu wenhua* 建筑与文化 2016 (5): 212–213. Print.

Zukin, Sharon. *The Culture of Cities*. Malden, MA: Blackwell, 1995. Print.

———. *Naked City: The Death and Life of Authentic Urban Spaces*. Oxford: Oxford University Press, 2010. Print.

Chapter 15

Rebuilding the Pavilion
"Doubled" Experience of Heritage at the Geo-Media Age

Xian Huang

INTRODUCTION

Recent decades have witnessed the rapid transformation of urban spaces and the high mobility of citizens in China. Demolition of old buildings and reconstruction are common in the regeneration process of cities, while the movement of people and goods within and between cities are much easier with the advanced transportation. The phenomena occur at the same time with the emergence of digital media technology, with which people are able to be connected in a global network in real time. As a space of flows is generated (Castells 2000), the high mobility of flows of information, people, and goods has also revealed its downside: its threat to traditional communities of geographical proximity. Physical location seems to be less important, because geographical distance is by no means a barrier of connection. People are experiencing a transformation of sense of identity with less stability and reliance to physical places (Bauman 2001), resulting in a lifted-out global information culture (Lash 2002).

The crisis of identity, conversely, is thought to push a culture of retroism that people try to maintain their internal spiritual balance by pursuing old-fashioned heritage for a sense of stability and safety in the unstable context brought by technological upsurge (Debray 2014). It is the fact that cultural heritage gains more attention in recent years to sustain the cultural identity of cities (Tiesdell et al. 1996). In China, more heritage sites are designated officially and more social groups take part in historic conservation, while the emerging geo-media is embedded into the conservative actions as well as people's everyday life.

The seemingly conflicting trends of high mobility and retroism catch my attention. Is the popularity of heritage merely a cultural spiral against the technological development? What is the relationship between people and heritage place? How do people experience the historical place in the current age? And how does new media technology affect the relationship? To answer the questions, I first review on the studies about the role of technologies in the relationship between human and the world.

TECHNOLOGIES AS MEDIATION OF
HUMAN–WORLD RELATIONSHIPS

Modern technologies, especially digital media, are usually blamed to be the cause of the identity crisis that destroys the balanced relation between human and the world. However, technologies are not the invaders of the preexisting human–world relationships, but always the mediation that enables the relationships (Verbeek 2005). Originating from the idea of Actor-Network Theory and post-phenomenology, Verbeek (2005) claims that technological mediation between humans and things is critical in the composition of social processes. Technological artifacts mediate the ways in which humans act and are present in the world and the ways in which reality is present to and experienced by humans. Certainly, it does not imply a technological determinism but emphasizes on the engagement of technologies in co-shaping the existence of human and the world. The materiality of technologies is highlighted here, on top of the common focuses on functions and meanings of the flows of information.

The recent emergence of geo-media, or spatial media, with the pervasive location-sensitive mobile devices and the mutual merging of location with digital content, enhances its role as a technological mediation that an always-mediated spatiality is produced in its mediation of socio-spatio-technical relations (Leszczynski 2015). Locations in material space still matter, and even become an "organization principle" of the new technology (Leszczynski 2015). The use of geo-media allows the revealing of a place and helps to locate oneself (Evans 2015), and the sense of locality is changed into "net locality" (Gordon and de Souza e Silva 2011) or "hyperlocality" (Rodgers 2018), with the coexistence and mutual construction of the two kinds of locality both physical and online, framed or mediated by digital technologies.

The relation between human and a heritage site, as a material space and a significant locale containing its cultural meanings, is not an exception but also mediated by technologies. Besides recreating a heritage in digital virtual forms, interactive digital media enable new modes of engagement for visitors and new forms of collective activities at the heritage site, and assist in returning people to a sense of place that involves more than simple location (Malpas 2008). The use of cross-media locative technologies, like GIS-related websites, generate a living heritage practice which further contributes to the social production of heritage and a new sense of connection, ownership and commitment toward a heritage place (Giaccardi and Palen 2008).

Heritage is, however, not only a spatial place with local meanings but also inherently carries the marks and memories derived from the previous time, providing temporalities of history and the past that people experience along with technological development. For technological philosophers, our senses of time, speed, and history are always involved with media technologies. Modern senses and ideas of time are constructed by the technology of production (Stiegler 2011), and the structures of our memory are varied with different technologies as carriers of memory (Stiegler 2009). In cultural studies, memories are seen to be produced from people's communicative practices naturally framed by technologies, and later be selected and stored in an institutional medium to construct a cultural memory, which can be transmitted for long ages and become the authorized history (Assmann 2011). The world of technology is not binary opposite to traditional or premodern culture (Bausinger 1990), but enables the transmission of our history as well as the construction of cultural heritage over time.

In digital age, the world is considered to be flattened into the present with flows of information without delayed reflection (Lash 2002). Speed and acceleration are seen to be emblematic of our modern times while the ubiquitous mediation of digital technologies results in multiple temporalities with heterogeneous rhythms and paces of everyday life, leading to new temporal experience and practice of human (Wajcman and Dodd 2017). For example, mediated memories in the digital age are continuously constructed and reconstructed with people's social practices of communicating and sharing, mixing the memories of both individual and collective, private and public (van Dijck 2007). Archives and time-logs on digital media which store documentary information of the past, in turn, present the temporal duration of life and promote a reflexive experience, in addition to the real time display of living practice (Pybus 2015; Brandtzaeg and Lüders 2018).

The ongoing construction of multiple temporalities of digital media enables a recreation of heritage. In this case, the recent retroism is not a return to local places and the past against technological progress, but a new sense of heritage, or a new heritage imaginary, constructed by digital technology and the related human practice. The current geo-media highlight their materiality with the portable devices embedded into everyday life, and their awareness of real-time location, which can further lead to a changing practice and experience at the heritage. Paying more attention to the temporal aspects, like history, memories, and archives, in the case of heritage and the emerging geo-media, I specify my research questions as follows: How do people practice on, sense and memorize the historical place mediated by geo-media? How do the memories and temporalities of a place (re)generate? What experiences of a heritage, both spatial place and temporal history, are (re)created in the current age?

The rebuilding of a pavilion in the Old City of Shantou (汕头), southeastern China, in 2016–2017, will be discussed in this paper, to look into the recent technological mediation of human–heritage relationships. Shantou has been a global trading port since 1860 and flourished as a commercial city. The pavilion, usually called the Pavilion of Small Park (小公园亭), is located at the center of the Old City, and serve as a public space for citizens and a landmark of the city since its first completion in 1934 (Figure 15.1). The governmental projects of renewing the Old City had been suspended for

Figure 15.1 Japanese Postcard of the Pavilion in 1940s. *Source*: Tan, Terence. *Memories of Old Swatow*. The Singapore Teochew Poit Ip Huay Kuan, 2011.

twelve years after punctuated in 2004, and restarted in 2016, beginning with another rebuilding of the pavilion. The rebuilding caught lots of social attention and participation. People's practices and ideas of the rebuilding both in materials and meanings are studied in my research, by applying the methods of participatory observation and in-depth interview. Historical documents and news reports on local media are also collected as secondary data.

MEMORIZING THE PLACE AND ITS REBUILDING

The large social attention paid to the rebuilding of the pavilion in late 2016 attributes to its important role as a representative historical place in the city. Located at the conjunction of five main streets in the Old City (Figure 15.2), in the center of the busiest commercial area but as an open public space, the pavilion used to serve large numbers of visitors with a place for rest and leisure. Especially, in the time of mid 1930s, when the city reached its economic peak, people from domestic areas and other countries rushed back and forth for trading, and the pavilion became a landmark where people gathered and encountered. Besides its location, the special architectural design, in a Chinese traditional style but with imported concrete materials, makes it more impressive among the surrounding arcade architectures with Western decorations.

Historical documents have shown that the pavilion was built in the name of commemorating President Sun Yat-sen. Besides these documents that are hardly accessible for public, stories and legends about the pavilion are spread orally in the local communities. It is said that the pavilion was, in fact, built

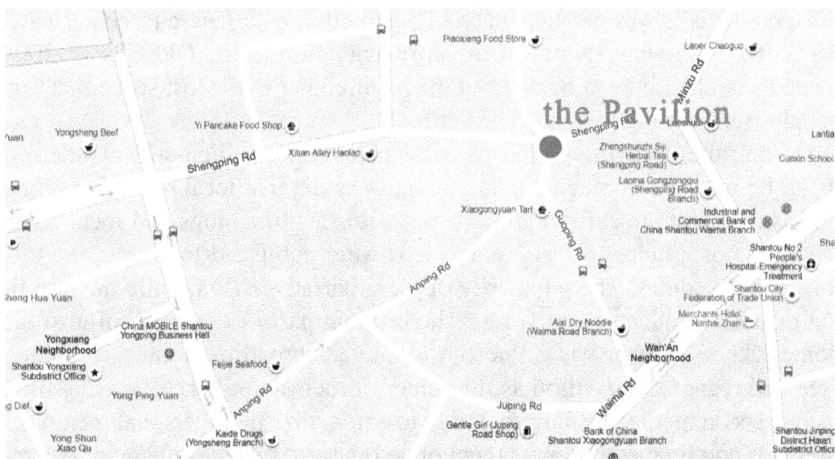

Figure 15.2 Map of the central Old City showing the location of the pavilion. *Source*: bing.com/maps.

to avoid the construction of another department store at the site as well as the possible commercial threat. The story implies the fierce competition among companies at that time, which is always a main topic of a business city. Other stories relate to the "Fengshui"[1] of the pavilion, that it is like a pearl in the middle surrounded by five streets like dragons. These stories are usually told to younger generations by elderly people who knew the stories from their parents aurally, and become known by more citizens, in and outside the local communities, recently when they are published in forms of textual articles on local newspapers and social media platforms.

The images of the pavilion, as a landmark, often appeared in the publications about the city in the 1930s, including journals and newspapers published by local intellectuals, as well as postcards and albums published by Japanese, Americans, and other foreigners. The place is depicted by photographing and painting in these publications, reproducing the scenes of the architecture as well as people in different dressing. The pictures, together with visual fragments of other sites in the city, create a general imagery of the busy city center. These publications were, on one hand, spread in the global trading network related to Shantou, promoting the urban imagery at that time. On the other hand, some of them have been kept by libraries and collectors, and also been digitized and displayed online open to more readers now. They become the most visible evidence of the urban history, based on the mediation of printed publications and their digital copies.

The oral stories and visible images build up most of the collective memories of the pavilion, especially after it was torn down in 1969 during the Cultural Revolution. The pavilion was rebuilt for the first time in 1997 after some revitalization of the city in economy, but the place was no longer as prosperous as it was because of the construction of a new city center eastward. The following failure of urban regeneration in the Old City in 2004 made the situation even worse, and the pavilion became a lost space that few people were willing to visit.

Despite the decaying of the physical place, the pavilion still maintained its high visibility on media, including mass media like local newspapers and televisions, and the early digital media platforms like blogs and local BBS, when lots of articles, photos, and videos were published to narrate the history and anecdotes. The popularity of these narratives is not only the nostalgia of local intellectuals who miss the brilliant past of the city, but also the homesickness of overseas Chinese who moved abroad from the surrounding area and regard the pavilion as the center of their home-place. Sharing their memories on media, the narrators tried to externalize their personal memories in words and images, and pass them on to others to make a collage of collective memories, which also offer some imaginary of the place to those who haven't been on the site.

The meanings of the pavilion, both historical and local, derive from people's memories of the past time. The memories, as shown above, are always mediated by media technologies, including printing and photographing, as the "new" media of the time when the pavilion was first constructed, and the recent technologies of digitization and Internet communication. The materialization of memories by media, in forms of printed photos and textual documents, and also of digital files based on the devices of computers and remote servers, enables them to be visible and shareable, and to preserve for long time regardless of the life span of a person and his ability of remembering. The preserved memory materials, further, are able to be spread to relevant persons globally and transmitted to future generations. Continuous memories of the place are thus created and maintained, despite of the discontinuity of the physical space with drastic transformations of demolition and rebuilding.

When it comes to the recent rebuilding in 2016, more natives, besides the intellectuals and pioneer users of the Internet, are able to publish their narratives and opinions, in texts, photos and videos, on social media platforms, thanks to the widespread use of Internet and mobile devices like smart phones. Social attention to the rebuilding became visible, and citizens were able to participate in practices on digital mobile media of recording, sharing, and discussing, as ways of experiencing and memorizing the transforming event of the place.

Not long after the release of the news, some citizens rushed to visit the place before the start of the demolishing project on October 25, 2016, and some of them later visited frequently during the process of construction. Taking photos was no longer a privilege of professional photographers, but became a habitual necessity whenever people visited a place. When taking photos, visitors watched the site on the screen mediated by lens when standing several meters away, rather than watching directly by eye. The changes resulted in a mediated experience of the place even when visitors were on site physically. The filters of lens, like angles, field sizes, depths of field, colors, etc., create and modify the visual presentation of the site flattened into fragmented 2D images, distracting the bodily sense of an overall built environment like smell and touch. Due to the embeddedness of the perspective of lens, photographers shifted their eyes back and forth between the screen and the site, and experienced the place in a more active but externalized way. On one hand, they moved their bodies to look for a better or alternative view, and used their arms and fingers for shooting the scene on phones and cameras. On the other hand, they relied on the material devices to catch and save the scene. The memories of visiting the place thus depend on their photographing practice and digital photos as memory objects. The most impressive details of the site are usually those that have been shot, even though they became part of the pictures coincidentally.

The photos also worked as a proof of physical visiting. Mobile geo-media like smart phones enabled visitors to post on social media platforms, like Weibo[2] and WeChat,[3] in the real time when they were at the pavilion. The posts were usually photos, texts, and videos about the construction with a locative check-in or tag showing the users' real-time location. When explaining the reasons of their visiting, photographing and posting, most visitors mentioned the historical and local meanings of the pavilion. They were enthusiastic to experience its rebuilding as a public event of the city. In addition, they did so because their family and friends did as well, and they tried to get involved with them in ways of posting and tagging. The real-time posts, joined by different citizens, repeated almost every day during the construction, and displayed the dynamic changes of the site online like live broadcasting. The continuous recording of the construction, containing fragmented individual narratives that were visible publicly, made a collage of collective memories as van Dijck (2007) has interpreted. The collage also worked as digital archives of the project, different from the archives of the reconstruction in 1997 that a retired professional photographer took photos every day voluntarily with his film camera and collected them into an album. In this time, the archives were produced with the participation of diverse citizens to photograph and shared on social media platforms, making an unplanned and unsystematic collection organized by thematic tags and locative check-ins.

The large amounts of imagery about the construction on social media were shared all over the world with the diaspora living in and out of China. Even though they were unable to be back home, the live broadcasting offered them a sense of participation in the local event. The news of the place was gained beyond the geographical barriers; some people living away even knew the latest situation earlier than some local residents. Comments, discussions and even debates happened in the related Weibo posts and WeChat groups, where diaspora were together with inhabitants sharing their opinions. While all reports on mass media spoke highly on the reconstruction, some critical ideas were expressed on social media about the necessity of rebuilding and the authenticity of the pavilion. The diverse opinions coexisted and dialogued online, resulting in hot topics on Weibo that gathered more social attention and participation. The online practice of discussion, generated a "net locality" framed by media technologies (Gordon and de Souza e Silva 2011), and further, contributed to future memories of the participators about the pavilion as a local and historical landmark.

The above practices of recording, sharing, and discussing are driven by people's memories and imagining of the previous pavilion, mediated by publications and oral stories including their contents and the material devices or bodies. Further memories of the place were, in turn, produced by the above practices both at the site and online, framed by the emerging geo-media.

With its characters of mobility, location sensitiveness and ubiquitous net-work connection in real time, geo-media has changed the way that people visit, experience and memorize a place and its event. People view the place via the screens, involve photographing and posting into the physical visit-ing, and keep their memories of the place by sharing and interacting online with others. Memories generated in this way are engaged with social practice and social relations partly online beyond the local place, but can be shared collectively with a sense of locality closely related to the physical site. New meanings of the place, both about spatiality and temporality, are then being constructed with the practices and memories in geo-media age, on top of the construction of material architecture.

REBUILDING THE MEMORIES AND IMAGINING OF HERITAGE

According to the official explanation, the rebuilding was to build a pavilion more resembling to the one in 1930s. In the first rebuilding in 1997, the construction was based on a collection of pictures. Besides the historical publications mentioned above, most of the images collected were taken not intentionally for architectural documentary, but scenery photos of the street views or documentaries of particular events that happened nearby. When there were no professional documents to be found to provide exact data of the previous building, the coincidental images with approximate looks played important roles in the reconstruction. Due to the lack of accurate data, the pavilion rebuilt in 1997 was obviously different from the former one both in size and in decoration.

Depending on the newly discovered original design diagrams, the local government and an enterprise-based investor decided to reproduce the origi-nal look of the pavilion by demolishing the 1997 one and building another one *in situ*. The diagrams on blueprint seemed to be much lighter and more fragile than the building constructed by concrete, but were ironically saved in the archival department while the material pavilion has been torn down into ruins. Professional information about the building designed more than eighty years ago was transmitted to the present, mediated by the symbolic numbers and lines on the diagrams and the material paper on which the diagrams were drawn. They were then digitized into digital images and sent to the designers before they were further redrawn and modified via com-puter programs. The historical archives, both specialized design plans and unintentional visual recordings like the previous collection of pictures, were thus combined with the current computer-aided designs, serving as the basic guidance of the reconstruction scheme. Through the mediation of paper and digital documents, a connection between the original pavilion and the recent construction project was built, promising a new pavilion resembled to the previous one.

It was believed by the constructors, the government and the investor, that the resemblance of buildings could provide similar senses of historical experience in the place. The resemblance largely referred to the similarity in visual style, rather than the authenticity of original materials and temporal traces. The emphasis on vision was related to the current focus on visual value in heritage conservation, especially when the historic building was regarded as a potential tourism destination. With the previous appearance as the representative imagery of local history, the new pavilion could provide a material landmark that natives and tourists could visit, view and consume. In addition, the look in the 1930s was considered prior to those in other ages, because it was the time when the city was in its heyday. The government, in the way of "restorative nostalgia" (Boym 2001) aiming at returning to a certain age, tried to reproduce the previous prosperous scenes of the Old City via the reconstruction of the pavilion, as the beginning of a larger restoration project of the area. Therefore, the rebuilding of the architecture was in fact a rebuilding of the imagining of a perfect past, based on the mediated historical documents and memories.

Despite the reference to the original professional data, the pavilion was inevitably a new one. Elements of the current time were included in the reconstruction. For example, construction materials were changed into woods rather than concrete, due to the investor as a timber merchant. Lighting with special design was also applied to provide more light for visitors in the evening but with limited interruption to its visual looks. Thus, the brand-new pavilion in a historical design offered multiple temporal senses for the visitors, both the past and the present, mediated by its material construction and the related symbolic meanings of time.

The completion of the reconstruction project was marked with a public performance of local opera in the new pavilion on January 9, 2017. It was a traditional ritual that the performance was offered to the "Gods of Land" to wish the place a good fortune. The unannounced performance attracted lots of passengers to make a stop. They sat on the benches in the pavilion, or stood around on the stairs and square, facing the performers in the central stage. Most people raised their arms to take photos and record videos with their smart phones. The images were posted on Weibo and WeChat shortly, and more citizens were attracted to come. People encountered and talked with each other when gathering together to watch the performance, generating an unexpected lively scene that was hardly seen there since the end of the 1990s. Chats in WeChat groups, participated by people from different places, also occurred when the photos and videos were shared in real time by people on site. In this case, geo-media like smart phones were involved for photo-taking and instant sharing, making the traditional performance as an attractive event for more visitors, in addition to an old-fashioned offering to gods for good wishes.

The pavilion, with the involvement of visitors' activities on site with geo-media, began to gain its popularity again, and to construct its contemporary meanings as a historical local place after the reconstruction of material space. The following Chinese New Year in February, 2017, witnessed a return of flows of visitors (Figure 15.3). According to governmental statistics, the number of visitors to the pavilion reached 300 thousand during the holiday.[4] Visitors, including citizens and diaspora who returned home for reunion, came here to have a look at the newly reconstructed building, with their memories and imagining of this representative landmark. Memory objects, including photographs about the historical street views and recent construction projects, traditional handicrafts made in local areas, and old machines like film projectors, were on display around the pavilion, introducing materialized memories into the space. QR codes that people could scan with their phones and be linked to digital archives of the place were printed on the exhibition boards. The future of the Old City was also exhibited with some computer-generated images about the proposed plans, to inform the visitors about the coming restoration project and ask for more suggestions. With the exhibits as archives of different time periods, visitors experienced the place

Figure 15.3   Social Activities at the pavilion during the Chinese New Year 2017. *Source:* Hongbin Zheng.

in a sense of temporality beyond the instantaneity. The past and the future were included in the real-time experience, when people encountered with the exhibits as technological mediation carrying temporal messages both on site and linked online, as well as the material building in a multitemporal look.

The meanings of the place were not only generated via the objects introduced by governments and other official institutions, but also via the spontaneous social practices of residents and other visitors. It was common for old neighbors and friends to meet coincidentally around the pavilion when the original residents gathered. They used to live nearby but haven't seen each other for long time after they moved away separately from the Old City. They greeted each other, and more importantly, friending on WeChat was usually included, as they were able to contact and follow others' moments later. The regathering of acquaintances recalled the collective memories of the place, and also generated new memories of the re-meeting itself. Worldwide social connection was also enabled with networked mobile geo-media. Posting, location checking-in, and chatting online were still active as it was during the construction. The real time scenes of the place were shared across the diaspora in other places via the connection of mobile Internet and social relations. Besides the global connection through social media platforms, some foreign elements were also introduced by local shop-owners to create some global atmosphere as it was in the imagined 1930s, as well as to attract more customers. A shop nearby changed its business from selling clothes into postcards and paintings about the Old City recently, and classical jazz from United States and Europe was played continuously during its opening hours. The shop-owner downloaded the music from a podcast application and played it on his smart phone connecting to a loudspeaker. He believed that the music was fitted to the architectural styles nearby and the imagery of the historical place as a global trading port. A sense of spatiality beyond the physical location was thus generated with the involvement of non-native music and worldwide social relations, mediated by geo-media technologies and the built environment in both local and foreign styles. People experienced the multiple spatiality, as well as the related temporality, of the place while interacting with material environment, both objects and other people, with their acts and bodily senses besides seeing.

The rebuilding of the pavilion, after the completion of the material building, continues to be processed for further meaning construction by mediums and social practices, to change the objective space into a historical local place, or a heritage. The new pavilion, as a "fake" antique, gains its real heritage meanings thanks to its location and exterior appearance, as well as the attached memories and imagining via exhibits and social activities. These objects and practices are mostly enabled and framed, or mediated, by geo-media in the current age, contributing the connection of different spaces, times, and people. In this case, geo-media as emerging technology is never a threat of a heritage, but a contributor to its construction, like what other technologies did as Bausinger (1990) depicts. Furthermore, sense of multiple times and spaces is generated during the experience of the heritage now with the embedded-ness of geo-media, as the meanings of the heritage place are not only those of the past and the local, but also those toward the future and the global. New memories and meanings are continuously created with the new practice on site and on geo-media, when the real-time posts about the local place are spread out worldwide and soon become digital archives. In this case, the rebuilding is not a return to a certain previous situation, but to produce a new history of the place and the city by introducing some modified elements of the past, as well as to create a new relation between people and the historical local place.

CONCLUSION: DOUBLED EXPERIENCE OF HERITAGE IN GEO-MEDIA AGE

The reconstruction of the pavilion contains a reforming of history, memories, place, and practice, both materials and symbolic meanings, with people's participation in ways of visiting, photographing, posting on social media plat-forms, chatting, image modifying, etc., embedded with geo-media. Multiple layers of spaces and times are included in this construction, and the multi-plicity is more appreciable in this age with the ubiquitous connection to other spaces and times via mobile real-time geo-media. People's experience of the pavilion, as a heritage place, in this case, is beyond the embodied world of here and now, but with the involvement of other contexts mediated by tech-nologies of both institutional media and other information-carrying materials.

I apply the word "doubled" in depicting the current experience of heritage both embodied and beyond embodied simultaneously. The word comes from Moores (2014) when he describes the intimate connection between online wayfaring, as movements through media settings, and embodied move-ments like those of fingers on keyboards and various touch-sensitive devices. While he focuses on the material habitual movements as the basis of online movements which are usually regarded as disembodied, I try to balance the emphasis on both physical and online practice by adding attention to the always-mediated connection of the embodied context, especially in the age when geo-media is embedded into our everyday life. Moreover, the "doubly digital" quality of contemporary media that Moores states mostly refers to the doubled-ness of spatial movements, but in my case of heritage, the doubled-ness of temporal senses is also included.

The doubled experience of heritage, or historical local place, is not a conjunction of the preexisting and separate worlds of on site and online, but a mutual construction of embodied context and the mediated one to co-shape the reality that is present to people. The mediated contexts, with multiple spaces and time, become part of the material reality in the embodied context on site in the real time, rather than virtual and imaginary worlds away from the scene. In terms of spatial aspect, the heritage place, embedded with information-carrying mediums, including building materials, exhibits, and geo-media like smart phones in individuals' hands, and with related practices, delivers a doubled sense of both local with its geographical location and indigenous imagery comparing to the other places, and global with its imported technological materials and out-reaching networks of symbolic content and social relations based on its physical location. In terms of temporal aspect, people's practices on heritage include the real time participating and recording during the construction process, as well as the mediated history learning and future creating. They generate not only the flows of instant actions and information, which Giaccardi and Palen (2008) state as a social production of heritage, but also archives with temporal depths which are stored in mediated ways both digital and printed, and accessible to the public in real time with geo-media. The rebuilding is, in this case, both a material construction of the historical building based on mediated multitemporal messages, and a meaning creation and archiving depending on the movements of materials/bodies and of information. Heritage inherently transmits a sense of history, with its historical meanings derived from its past and represented by its materials and related narratives, and is now experienced also as a recent construction containing contemporary materials and meanings, which is continuously renewed by people's ongoing practices based on their previous experiences and memories. Multiple temporalities are blended, rather than flattened as Lash (2002) claims, in the doubled worlds, sustaining the ever-lasting significance of the embodied historical place which connects history, today and future.

The doubled-ness of either space or time is not unique in the geo-media age, but always exists thanks to technological mediation between human and the world, or in my case, the historical local place. Ever since the very beginning, the pavilion offered doubled experiences in space of both local and global, connecting with overseas Chinese and trading networks, and in time of both real time and long-lasting with business stories and urban history. The introduction of geo-media enhances the doubled-ness with its characters of location sensitiveness, real-time feedback, ubiquitous connection, and digital archiving. The location sensitiveness of the device enables its embeddedness into embodied space; the real-time feedback allows the instant practice and reaction; the ubiquitous connection offers worldwide networked links beyond the physical location; and the digital archiving serves long-lasting storage of information and practice. The combination of these material and informational characters blurs the boundary between the embodied context of here and now, and the mediated/online context of other spaces and times, as real-time connection with the outward places and the past histories are both accessible. With geo-media, the shuttling of the two is less conscious but more habitual in people's everyday practice, contributing to the integration and mutual construction of embodied and mediated experience, or doubled experience.

The new experience is not merely driven by technologies, but based on the interaction between human and geo-media. On one hand, some mediated practices, like photographing and posting on social media platforms, are more and more bodily habitual in one's everyday life. These practices offer experiences of a historical local place by practical knowing and by habitation of online sharing and socialization, sometimes without cognitive reasoning and purposeful aims as historical documentation. On the other hand, intentional and rational practices, or actions as Verbeek (2005) mentions, of designing, documenting, and discussing are still included in the participation of rebuilding event. Different actors, such as local government, residents, and diaspora, engage in the construction of both materials and meanings when applying information and practices mediated by geo-media into the embodied context.

The doubled experience, with increasing shuttling and mutual construction of the embodied and mediated contexts, results in a sense of the world as unstable, ambiguous, and ever-evolving, that it is continuously reconstructed by media technologies and human's practice. Progressive sense of space has been discussed by geographers like Massey (1997), stating the mutual construction of globality and locality. A progressive sense of time can be added here, to highlight the interaction of real-time-ness and duration. With the doubled experience and the related progressive sense, the governmental project of rebuilding the pavilion for "restorative nostalgia" has thus been changed into a socially engaged "reflective nostalgia" (Boym 2001) to create a better future with the recollection of shattered fragments of memory and temporalized spaces. Heritage, accordingly, is not necessarily old-fashioned and solidified for indigenous communities, nor something people pursue backward against the technological development, but open to further (re) construction when exotic and new technologies and human practices are introduced and keep transforming for future generations. With the progressive sense of heritage, the conflict and negotiation between technological transformation and retroism is no longer a problem, and new understanding of history, memory and humanity is waiting for further exploration considering the emerging technological mediation of geo-media.

NOTES

1. Fengshui is an ancient system of China about the relation between spatial arrangement and good fortune of the places.

2. Weibo is a microblogging website and mobile application developed by Sina Corporation, one of the biggest social media platforms in China.

3. WeChat is a multipurpose messaging and social media application developed by Tencent Corporation, also one of the most popular platforms in China.

4. See "Summary of Spring Festival Holiday Tourism in Shantou City". *Governmental Website of the Provincial Tourism Administration in Guangdong*, February 9, 2017, http://www.gdta.gov.cn/gzdt/dsdt/38463.html. Accessed February 28, 2019.

WORKS CITED

Assmann, Jan. *Cultural Memory and Early Civilization: Writing, Remembrance, and Political Imagination*. Cambridge: Cambridge University Press, 2011.

Bauman, Zygmunt. *Community: Seeking Safety in an Insecure World*. Cambridge: Polity Press, 2001.

Bausinger, Hermann. *Folk Culture in a World of Technology*. Translated by Elke Dettmer. Bloomington: Indiana University Press, 1990.

Boym, Svetlana. *The Future of Nostalgia*. New York: Basic Books, 2001.

Brandtzaeg, Petter B. and Marika Lüders. "Time Collapse in Social Media: Extending the Context Collapse." *Social Media + Society* January–March (2018): 1–10.

Castells, Manuel. *The Rise of the Network Society* (2nd Version). Oxford: Blackwell, 2000.

Debray, Regis. *Introduction of Mediology* (Chinese Version). Translated by Wenling Liu and Weixing Chen. Beijing: Communication University of China Press, 2014.

van Dijck, Jose. *Mediated Memories in the Digital Age*. Stanford: Stanford University Press, 2007.

Evans, Leighton. *Locative Social Media: Place in the Digital Age*. London: Palgrave Macmillan, 2015.

Giaccardi, Elisa and Leysia Palen. "The Social Production of Heritage through Cross-media Interaction: Making Place for Place-making." *International Journal of Heritage Studies* 14.3 (2008): 281–297.

Gordon, Eric and Adriana de Souza e Silva. *Net Locality: Why Location Matters in a Networked World*. Oxford: Blackwell, 2011.

Lash, Scott. *Critique of Information*. London: SAGE, 2002.

Leszczynski, Agnieszka. "Spatial Media/tion." *Progress in Human Geography* 39.6 (2015): 729–751.

Malpas, Jeffery E. "New Media, Cultural Heritage and the Sense of Place: Mapping the Conceptual Ground." *International Journal of Heritage Studies* 14.3 (2008): 197–209.

Pybus, Jennifer. "Accumulating Affect: Social Networks and Their Archives of Feelings." *Networked Affect*, edited by Ken Hillis, et al. London: MIT Press, 2015. 235–250.

Rodgers, Scott. "Roots and Fields: Excursions Through Place, Space, and Local in Hyperlocal Media." *Media, Culture & Society* 40.6 (2018): 856–874.

Stiegler, Bernard. *Technics and Time, 2: Disorientation*. Translated by Steven Barker. Stanford: Stanford University Press, 2009.

Stiegler, Bernard. *Technics and Time, 3: Cinematic Time and the Question of Malaise*. Translated by Erping Fang. Stanford: Stanford University Press, 2011.

Tiesdell, Steven, et al. *Revitalizing Historic Urban Quarters*. London: Butterworth-Architecture, 1996.

Verbeek, Peter-Paul. "Artifacts and Attachment: A Post-script Philosophy of Mediation." *Inside the Politics of Technology: Agency and Normativity in the Co-Production of Technology and Society*, edited by Hans Harbers. Amsterdam: Amsterdam University Press, 2005. 126–146.

Wajcman, Judy and Nigel Dodd. *The Sociology of Speed: Digital, Organizational and Social Temporalities*. Oxford: Oxford University Press, 2017.

Conclusion

Xinmin Liu and Peter I-min Huang

Inspired by the keen insights of part III in this volume, it is only logical and appropriate to follow up by reiterating the key words highlighted in the Introduction as well as the entire volume: "Enact the human–nature bond bodily!" As a closure to our critique of the disembodied viewer of natural beauty, we would like to direct our focus on Ban Wang (one of the contributors to this volume); we owe it to Ban Wang for boosting our approach with a critical and impactful strategy to delink "the aesthetic" (as *felt* sentiments) from "aesthetics" (as *fetishized* knowledge), as he has traced the evolving trajectory of Chinese aesthetics to its origin of the Enlightenment Age in the West—his present chapter also resonates with his impactful endeavor. By setting the two concepts apart, Wang returns the enactment of bodily reaction in appreciating natural beauty to the front and center of our aesthetic response and returns the embodied experience to the fertile grounds of human development, politics, ethics, and social life, thus paving the way for the ensuing advocacy of the embodied and embedded mode of human co-existence with the natural world. In his essay named "Use in Uselessness: How Western Aesthetics Made Chinese Literature More Political," Wang assails the wrongheaded drive led by some theorists to hoist the human aesthetic activities to the abstract, institutional pedestal by stripping them of their corporeal, societal, and cultural ties, and he defines the jaded subjectivity of such aesthetics "as a thinly veiled autonomous realm of appreciating beauty"—fittingly reminiscent of scholar Fan's faulty claim we discussed in Introduction (4). As we recall, Fan's assertion for aesthetic autonomy makes it virtually infeasible for environmental scholars to play an active role to contest and mediate the divide between agrarian and industrial values of the past and present, between techno-science and aesthetic activity; while quoting Xu Fuguan 徐复观, Fan states that "if you respect and love nature she will reward you a thousand

267

times. She won't make you feel pressured; on the contrary, she 'liberates' and 'comforts' your spirit" (Fan 270). Fan then cites the example of longevity many renowned painters in Imperial China enjoyed, illustrating how nature rewarded them by letting them "[benefit] from beautiful landscapes in their paintings" (270). It is regrettable that Fan has omitted the reality of the social, economic, and cultural privileges that permitted the elite literati to lead their lives in peace, leisure, and comfort. On the other hand, it is extremely mis-leading for Fan to declare that the ideals of "liberating" and "comforting" can only be fulfilled spiritually in one's imagined world by looking at his painted images of nature, and nowhere else!

In reviewing the rift dividing science and green thought, the renowned ecocritic Ursula Heise heartily upholds scientific knowledge of nature as "one of the cornerstones of ecocriticism," yet she regards it as "a place where different visions of nature and varying images of science, each with their cultural and political implications, are played out" (ASLE website 2). By interacting and contesting at this place, she writes, "it also allows the critic to see where literature deviates—or, in some cases, wishes or attempts to devi-ate—from the scientific approach in view of particular aesthetic and ideologi-cal goals" (ASLE website 3).[1] Ban Wang has done exactly that by revealing that the disembodied aesthetics in Chinese art and literature has effectively brought into being a modern extension of aesthetics soon to be deployed as a techno-scientific makeover by China's modernists, and it was done through "compartmentaliz(ing) organic, interconnected human activity on the model of commodity production and specialization" (3). In his ensuing eco-critical writings, Wang has persistently engaged in uncovering the role of complicity assumed by followers of techno-scientific triumphalism prevalent in early modern China. He is among the earliest of the Chinese ecocritics with a cos-mopolitan vision to uncover the derivative ties between imported notions of European aesthetics and the wholesale westernization of techno-sciences in late-Qing China known as the "Movement of Western Affairs" (洋务运动); his chastisement of the partial viewpoint of pushing for environmental justice to the neglect of social justice is spot on in anticipating the repeated rallying calls for "social justice as *environmental justice*" to the swelling ranks of the environmental humanities. By harking on the virtues of the "Great World Community" boldly envisioned by the late-Qing Confucian reformer Kang Youwei (康有为), Wang aims the upshot of his ecocritical writings at meting out an antidote for treading slavishly the footsteps of the Western modernity of industrialization and market-driven ventures and, alternatively, for foster-ing an ecocritical sensibility that is true of and dedicated to China's deep-rooted ecological wisdom.

Spurred on by kindred-spirited ecocritics like Wang and Heise, we have taken issue with the correlation between human existence and the natural world, maintained vigilance against the destructive and exploitative modes of disembodied human thinking toward the natural lifeworld, and remained committed to the ideal of "human–land affinity" (人地亲和). The issues we have explored in the proceeding chapters bear on an area intersecting aesthetics, ethics and cultural and sociological studies, and at the heart of our inquiry are these questions: What can be done, in an age saddled with global spread of killer pandemics such as COVID-19, to enhance our faith in land–human affinity and seek to restore and rebuild a livable and sustainable habitat on Earth for both humans and nonhumans? How can we rejuvenate a constructive mentality of responsibility and empathy toward the natural world while being consumed by the debilitating milieu of Anthropocene? We have taken a modest step forward in search of answers that should help us avert appreciating the values of the natural world with diverse forms of species as if we were all leisure-seeking viewers strolling casually through a picture gallery, a typical human comfort zone formed over centuries with a price tag of approaching life endangerment for all earthlings.

There has been heartening evidence in East Asia lately to attest to the fact that endeavors like this volume have slowly yet steadily brought our common cause to fruition. From a swarth of impactful cases, we select and share with our readers one of impressive ingenuity and bountiful bliss (pun intended) that is meaningful in more ways than one.[2] The case involves a cluster of renovated rural houses known locally as *Yangjiale* 洋家乐 (literally translated as "dwellings of foreign bliss"), which were once traditional houses nestled along small brooks flowing through pristine and lushly shaded gullies off the side of the well-known resort town *Moganshan* 莫干山 in Zhejiang Province, East China. These farm houses were first sublet to a group of foreign entrepreneurs working in Shanghai-based finance and business corporations in 2007. Led by a South African named Grant Horsfield, the group started subletting and remodeling these largely run-down traditional houses and turning them into livable lodges for their bicycling trips away from Shanghai over weekends. Their idea became viral overnight and the houses opened to enthusiastic throngs of travelers from across China and even overseas. In 2011, Horsfield was joined by his Hong Kong architect wife, Delphine Yip, and together they renovated the nearby stables for farm animals; by blending cutting-edge technological design with traditional methods for rural house-building (such as houses with rammed-earth walls), they succeeded in building a top-line resort area named "Naked Retreats" (*Luoxingu* 裸心谷) catering mostly to eco-friendly travelers and were awarded as China's first globally recognized resort in sustainable tourism. Intriguingly, it has in part to do with the name "naked heart" (*luoxin* 裸心). Evidently, being naked in this context has nothing to do with human nudity, but everything to do with the human resolve to free one's heart from all its trappings—to leave all one's desires and possessions behind in the city and be exposed to the natural world here solely and unreservedly!

It is important to note the term "Naked Retreats" implies that we commit to a voluntary *stripping* ourselves of all the comfort, convenience, and leisure we have grown accustomed to while living in our urban "comfort zones"; and it entails a willing surrender of our bodily being to activities that force us to oftentimes be on the move outdoors and coping with simple living conditions indoors. It also signals one's acceptance of being directly exposed to the erratic turns of the local weather, being constantly mobile and physically assertive in body and mind; while doing so, we consume as little natural resources and leave as few carbon footprints as possible. It compels us to embrace a solution of "beating a retreat" from the tedium and hassle of the high-tech world in urban centers, not just by means of occasional escapades to rural places in search of peace and quietude, but by way of robustly retooling our physical conditions and reinvigorating our neglected and "underrated" sensory functions, that is, smelling, hearing, touching, and tasting; in doing so, we can regain a sharper alertness and reconnect to the "raw ambience"— the circulating flow of the vital *Qi*—shared with the natural world in which humans are expected to partake of a more holistic environment of health and stability. To be clear, our use of the term "raw ambience" echoes with the Daoist transversal being and the Buddhist nonbeing, but it resonates even more with the kind of *ensemble effect* in which the human brain interacts not only with all other organs of the human body, but through them, the organismic pool of the biophysical world that impacts incisively on the human mind.

Taking cue from Horsfield's unique approach to eco-travel, the local townsfolks have now begun to turn their traditional farmhouses into low-carbon, little-waste guesthouses for travelers and weekend vacationers from nearby cities who are attracted here by its idyllic natural scenery and low travel costs. In the meantime, the more culture-minded local residents have revived and embellished the local ambience by means of the splendid local cultural heritage—indeed it helps to breathe life into the folklore of certain *genius loci* well-stocked with rich literary output and cultural abundance. They surely pushed, if not directly ignited, a ground swell in promoting ecotourism in the rural regions of East Asia. Nothing speaks more eloquently than this as a living and thriving example of Wei-min Tu's notion of the Continuity of Being that ties humans with the natural world closely and infinitely.

NOTES

1. Ursula K. Heise, "Science and Ecocriticism," *The American Book Review* (Victoria, TX: The University of Houston-Victoria), 18.5 (July–August 1997): 4.

The current version is a permitted posting of Heise's book review carried by The Association of Studies of Literature and Environment at its website: https://www .asle.org

2. The following passages have been adapted by Xinmin Liu from his essay entitled: "Foreword: Placing the Public(s) in Revitalizing Local and Rural Cultural Heritages" published by *Communication and The Public*, 3.4 (December 2018). Permission has been obtained from the CAP journal for Liu's adapted passages.

WORKS CITED

Fan, Meijun. "Ecological Consciousness in Traditional Chinese Aesthetics," Trans. Yang Fengzhen, *Educational Philosophy and Theory*. Philosophy of Education Society Australasia. Vol. 33, No. 2, 2001. DOI: 10.1080/00131850120040618.4+

Wang, Ban. "17. Use in Uselessness: How Western Aesthetics Made Chinese Literature More Political," *A Companion to Modern Chinese Literature*. Ed. Yingjing Zhang, 2016. 1–17. DOI: 10.1111/b.9781118451625.2016.00018.x.

Index

About the Editors and Contributors

EDITORS

Xinmin Liu is an associate professor of Chinese and Comparative Cultures in the School of Languages, Cultures, and Race at the Washington State University. He received his Ph.D. in Comparative Literature at Yale in 1997, and his teaching and research has since been chiefly cross-cultural and inter-disciplinary, dealing with culture, ethics, and social thought. His published works include *Signposts of Self-Realization: Evolution, Ethics and Sociality in Modern Chinese Literature and Film* with Brill Academic Publishing (2004) and many journal articles and book chapters on literary and cultural themes in modern China. Since 2005, he has been focusing on cultural geography, landscape aesthetic, and ecocriticism in China and the West and has edited a few special issues on ecocritical studies of art, literature, and culture for journals such as *Forum for World Literature Studies* (China), *Tamkang Review* (Taiwan), and *Communication and The Public* (US). He is currently completing a monograph entitled *The Return of the Agrarian Imaginary: Interlaced Agencies in Revitalizing Human-Land Affinity*.

Peter I-min Huang is a professor emeritus at Tamkang University. His areas of interest include English and Chinese Literature, ecofeminism, ecopoetry, indigenous studies, postcolonial ecocriticism, posthumanism, science fiction, and climate fiction. Dr. Huang is a founding member of ASLE-Taiwan. He served as the Tamkang University English Department Chair for two terms (2007–2011). During this same period of time, he was the conference organizing chairperson for The Fourth Tamkang International Conference on Ecological Discourse (May 23–24, 2008) and The Fifth Tamkang International Conference on Ecological Discourse (December 17–18, 2010).

Journal articles by him include publications in *Neohelicon*; *Journal of Poyang Lake*; *CLCWeb: Comparative Literature and Culture*; *World Literature*; and *Foreign Literature Studies*. Book chapters include chapters in *Transecology: Transgender Perspectives on Environment and Nature* (2020); *Literature and Ecofeminism: Intersectional and International Voices* (2018); *Ecocriticism in Taiwan: Identity, Environment, and the Arts* (2016); and *East Asian Ecocriticisms: A Critical Reader* (2013). He is the author of *Linda Hogan and Contemporary Taiwanese Writers: An Ecocritical Study of Indigeneities and Environment* (Lexington Books, 2016).

CONTRIBUTORS

Kathryn Yalan Chang is an associate professor in English Department at National Taitung University. Her current research interests include nature therapy, material ecocriticism, environmental humanities, environmental justice and activism, food and animal studies, and ecofeminism. Chang's research project was entitled "Affective Ecocriticism: Nature, Therapy, and Science." Her latest publications include: "After the Japan 3/11 Disaster: Slow Violence and Slow Living," in *Tamkang Review* 47.2 (June 2017), "'Slowness' in the Anthropocene: Ecological Medicine in *Refuge* and *God's Hotel*," *Neohelicon* (2017), and "A Material Feminist reading of *The Awakening* and *The Yellow Wallpaper*," *Review of English and American literature* (2018).

Kiu-wai Chu is an assistant professor in environmental humanities and Chinese studies at Nanyang Technological University, Singapore. He earned his PhD in Comparative Literature in University of Hong Kong, and his previous degrees from SOAS University of London and University of Cambridge. He was a visiting Fulbright scholar in the University of Idaho, and postdoctoral fellow in the University of Zurich and Western Sydney University. His research focuses on ecocriticism, environmental humanities, and contemporary cinema and visual art, specifically in Chinese and Southeast Asian contexts. His works have appeared in *Transnational Ecocinema*; *Journal of Chinese Cinemas, Asian Cinema*, and *Chinese Environmental Humanities*.

Simon Estok is a professor and senior research fellow at Sungkyunkwan University (South Korea's first and oldest university). Estok teaches literary theory, ecocriticism, and Shakespearean literature. His award-winning book *Ecocriticism and Shakespeare: Reading Ecophobia* appeared in 2011 (reprinted 2014), and he is a co-editor of five books: *Anthropocene Ecologies of Food* (June 2021), *Mushroom Clouds: Ecological Approaches* *to Militarization and the Environment in East Asia* (March 2021), *Landscape, Seascape, and the Eco-Spatial Imagination* (2016), *International Perspectives in Feminist Ecocriticism* (2013), and *East Asian Ecocriticisms* (2013). His latest book is the much anticipated *The Ecophobia Hypothesis* (2018; reprinted with errata as paperback in 2020). Estok has published extensively on ecocriticism and Shakespeare in journals such as *Mosaic, Configurations, English Studies in Canada, PMLA*, and others. He is currently working on a book about slime in the Western cultural and literary imagination.

Xian Huang is a lecturer in the Cheung Kong School of Journalism and Communication at Shantou University, and a research member in the Center of Information and Communication Studies at Fudan University, China. She received her doctoral degree of communication in the School of Journalism at Fudan University. Her main interests of research include urban communication, cultural heritage and digital media. Her publications have appeared in *Communication and the Public* (2018), *Review of Communication Studies in China: The Turn of Media* (2019), *Journalism and Mass Communication Monthly* (2020), and *News Research* (2020). She has been working on the case of the Old City of Shantou, Guangdong Province, China, since 2012, and focused on how digital media change the relation between citizens and the historical place in her doctoral thesis.

Young-Hyun Lee teaches at Sungkyunkwan University in Seoul. Her publications of translated books include *Do You Know The "Comfort Women" of the Imperial Japanese Military?* (2017) and 나의 첫여름 (My First Summer in the Sierra), which Dr. Lee co-translated with Won-Chung Kim in 2008. Lee recently received two one-year research awards from the National Research Foundation, which were completed at Kangwon University for the academic years 2018–2019 and 2019–2020. She is the author of such scholarly articles as "Trans-corporeality, climate change, and *My Year of Meats*," *Neohelicon* 47 (A&HCI, 2020), "Food Transformation Technology in Paolo Bacigalupi's *The Windup Girl* and *What it Means for Us*," *Kritika Kultura* 33/34 (A&HCI, 2019), and "The Different Representations of Suffering in the two versions of *The Vegetarian*," *CLCWeb* 21.5 (A&HCI, 2019).

Hua Li is a professor of Chinese and coordinator of Chinese program in the Department of Modern Languages at Montana State University. She received her doctoral degree in Asian Studies from University of British Columbia, and her primary research field is modern and contemporary Chinese literature. She published her monograph *Contemporary Chinese Fiction by Su Tong and Yu Hua: Coming of Age in Troubled Times* in 2011, and has published refereed journal articles and book chapters on various topics in contemporary Chinese literary fiction, cinema, animation, and science fiction in *Cambridge History of Science Fiction, Science Fiction Studies, China Information, Communication and the Public, Journal of Chinese Cinemas, Frontier of Literary Studies in China, Forum for World Literature Studies,* and other peer-reviewed journals. Her second book on Chinese science fiction in the 1980s—*Chinese Science Fiction during the Post-Mao Cultural Thaw*— is forthcoming (2021).

Jialuan Li is a lecturer at the Shanghai University of Engineering Science. He graduated in 2019 from the English language and literature PhD program at Nanjing Normal University. His research area is generally in ecocriticism, but he also has academic interest in Chinese classics, especially Daoist philosophy, and the comparative study of ecocriticism and Daoist ecology. He published a co-authored book, *Ecofeminism*, with his PhD supervisor, and several CSSCI (Chinese Social Sciences Citation Index) articles on ecocriticism and Daoist philosophy.

Xin Ning is an associate professor in Comparative and World Literature Program, College of Humanities, Jilin University. His fields of interest include modern Chinese literature and culture, Anglo-American modernism, eco-criticism, and translation studies. His latest book *zhuti yu shenghuo shijie: xiandai zhongguo wenxue zhong de ziran, shengtai yu ren* (Subject and Modern World: Nature, Ecology and the Human Figure in Modern Chinese Literature) was published in 2017 by Jilin University Press.

Kenichi Noda is a professor emeritus of Rikkyo University in Tokyo, Japan. Noda was the first president of the ASLE-Japan when it was established in 1994. His expertise lies in American literature, nature writing, and environmental literature, and his published work covers American writers like Emerson, Thoreau, Dickinson, Abbey, Dillard, Lopez, Nelson, T. T. Williams, as well as Japanese environmental writers. His scholarly interest mainly focuses upon the "postromantic" attitude toward nature found in the contemporary writers. Especially, Post-Romanticism, Land/Landscape dichotomy, and Alterity Turn are among the keywords that he has emphasized and been concerned with. His published books include *Correspondence and Representation: The Nature of Nature Writing* (2003) and *It Is We Who Are Lost: Nature, Silence, and the Others* (2016). He is the translator of Edward Abbey's *The Journey Home*. Among his many articles are "Textualized and De-textualized Nature: Emerson, Thoreau, and Abbey" (2003), "A Wild Dog Chase: The Represented Wildness in Fujiwara Shinya's Works" (2003), "The Absence of Here/Now: *Walden* as Discovery Narrative" (2006), "Princess Mononoke and the Language of the Wild: The Alterity Turn in the View of

Nature" (2016), "Great Nature's Almanac: The Alterity Turn in the Works of Ishimure Michiko" (2018).

Iris Ralph is a professor in the English Department of Tamkang University. She specializes in ecocriticism, animal studies, plant studies, and posthumanism. Ralph's most recent journal articles include publications in *Kritika Kultura*; *ISLE: Interdisciplinary Studies in Literature and Environment*; *ICL: International Comparative Literature: A Journal Devoted to Research in Transcultural Encounters*; *Neohelicon*; *Tamkang Review*; *AJE: Australasian Journal of Ecocriticism and Cultural Ecology*; and *William Carlos Williams Review*. Book chapters include publications in *Doing English in Asia* (2016), *Ecocriticism in Taiwan* (2016), and *Dystopias and Utopias on Earth and Beyond* (2021). She is the author of *Packing Death in Australian Literature: Ecocides and Eco-Sides* (2020).

Scott Slovic is the University Distinguished Professor at the University of Idaho, USA. He was the founding president of the Association for the Study of Literature and Environment (ASLE) from 1992 to 1995. He also edited ASLE's journal, *ISLE*, for twenty-five years until 2020. Slovic is the author and editor of numerous books in the environmental humanities and has lectured and taught widely in mainland China, Japan, South Korea, Taiwan, and throughout the world. He currently co-edits the book series Routledge Studies in World Literatures and the Environment and Routledge Environmental Humanities.

Lili Song is an associate professor at Tsinghua University in Beijing, China. Educated at Northeastern Normal University (majoring in American Literature) and Beijing Language and Culture University (Ph.D., in comparative literature 2005), she taught in a few colleges in China's northeastern region from 1989 to 1995, and has been teaching at the Department of Foreign Languages and Culture at Tsinghua University since 1998. She has also worked at Northwestern, Georgia Tech, and Baylor as a visiting scholar in the United States. She began her study in ecocriticism with her doctoral dissertation titled "On the Construction of Literary Ecology." She has published widely in Chinese journals and contributed to *ISLE* with critical essays such as "On the Eco-niche of Literature," "The Vision of Literary Criticism toward Nature," "To Be or Not to Be: On the Pathos of Chinese Environmental Writing about the Yellow River."

Ban Wang is the William Haas Endowed Chair Professor in Chinese Studies in the Departments of East Asian Languages and Cultures and Comparative Literature at Stanford University. He was the Yangtze River Chair Professor at East China Normal University in Shanghai. His major publications include *The Sublime Figure of History: Aesthetics and Politics in Twentieth –Century China* (1997), *Illuminations from the Past* (2004), *History and Memory* (in Chinese, 2004), and *Narrative Perspective and Irony in Chinese and American Fiction* (2002). His book *China in the World: Culture, Politics, and World Vision* is forthcoming from Duke UP, 2022. He edited *Words and Their Stories: Essays on the Language of the Chinese Revolution* (2010) and *Chinese Visions of World Order* (2017). He co-edited *Trauma and Cinema* (2004), *The Image of China in the American Classroom* (2005), *China and New Left Visions* (Lexington Books, 2012), and *Debating Socialist Legacy in China* (2014). He was a research fellow with the National Endowment for the Humanities in 2000 and at the Institute for Advanced Study at Princeton in 2007. He has taught at Beijing Foreign Studies University, SUNY-Stony Brook, Harvard University, Rutgers University, the University of Zurich, Seoul National University, Yonsei University, and E. China Normal University.

David Wang is a professor of architecture emeritus at Washington State University, where he taught architectural theory for many years. His theoretical interests include retrieving ancient philosophical worldviews, both East and West, and applying them to current issues in material culture. This research has resulted in the books *A Philosophy of Chinese Architecture Past Present Future* (2017) and *Architecture and Sacrament: A Critical Theory* (2020). Dr. Wang has also published extensively on design research, and this has allowed him to lecture widely in China, the Scandinavian countries, and the United States. He is a co-author of *Architectural Research Methods* (second edition, 2013) and co-editor of *Research Methods for Interior Design: Applying Interiority* (2020).

Qingqi Wei is a professor of English at Southeast University, China. He is one of the earliest promoters of ecocriticism and ecofeminism in China, and has published articles on domestic and international journals including *Interdisciplinary Studies of Literature and Environment*. Wei is also a well-known translator with focus on classics of English languages and ecocritical monographs.

Philip F. Williams is an actively teaching professor of East Asian studies in the Department of Modern Languages and Literatures at Montana State University. He is also the professor emeritus of Chinese at Arizona State University, where he was promoted to full professor in 2000 and received the ASU Humanities Research Award in 1990. Since the mid-1980s, Williams has taught a wide variety of courses in East Asian languages and cultures at nine different research universities in North America, China, and Australasia. His research on East Asian languages and cultures has resulted in more than ten books and over 200 shorter publications such as refereed journal articles, scholarly book chapters, essays, and translations. Publication highlights include: editor and co-translator of Hú Píng's *The Thought Remolding Campaign of the Chinese Communist Party-state* (2012); co-editor of the festschrift for the prolific literary translator Burton Watson (2015); author of the peer-reviewed article "Fundamentalist Thinking in Chinese Maoist 'Thought Remolding'" (in *Asian and Asian-American Philosophers and Philosophies* 16.2 (2017)); and author of a book chapter in *Routledge Handbook of Modern Chinese Literature* entitled "Three Independent Chinese Novelists: Shěn Cóngwén, Xǔ Dìshān, and Qián Zhōngshū" (2019).

Sijia Yao teaches Chinese literature, film, media, and language in the University of Nebraska, Lincoln. She graduated from Purdue University with a PhD in Comparative Literature. Her writings on comparative literature and Chinese studies have appeared in such peer-reviewed journals as *The Comparatist, Comparative Literature Studies, Forum for World Literature Studies, Rocky Mountain Review, Tamkang Review,* and *Movie Review.* She has presented papers in prestigious conferences and workshops, such as MLA convention, ACLA convention, and Taiwan Studies Workshop. She is currently working on a book project: *Cosmopolitan Love: Affect as Engagement in D. H. Lawrence and Eileen Chang.*

www.ingramcontent.com/pod-product-compliance
Lightning Source LLC
Chambersburg PA
CBHW050630280326
41932CB00015B/2592